T0298357

Implementing Energy Efficiency in Industries

This book focuses on designing, implementing, and verifying the performance of energy efficiency and conservation (EE&C) projects in relevant industries from a practitioner's perspective. Various techniques and approaches are presented using case studies collated from the author's notes from about four decades of working in process industries and two decades as an international sustainable energy consultant.

Features:

- Provides a broad overview of the main issues in implementing energy efficiency in industries.
- Focuses on implementation issues – technical, financial and employee engagement.
- Provides a brief description on the fundamental thermodynamic principles that drive efficiency and conservation.
- Includes a comparative evaluation of ESCO performance contract implementation.
- Charts out the energy efficiency journey, developing long-term goals and short-term activities.
- Includes case studies related to energy efficiency in energy-intensive large industries.

This book is aimed at professionals working in energy engineering, industry, efficiency, and policy.

Implementing Energy Efficiency in Industries

G C Datta Roy

CRC Press
Taylor & Francis Group
Boca Raton London New York

CRC Press is an imprint of the
Taylor & Francis Group, an **informa** business

Designed cover image: www.shutterstock.com

First edition published 2025
by CRC Press
2385 NW Executive Center Drive, Suite 320, Boca Raton FL 33431

and by CRC Press
4 Park Square, Milton Park, Abingdon, Oxon, OX14 4RN

CRC Press is an imprint of Taylor & Francis Group, LLC

© 2025 G C Datta Roy

Library of Congress Cataloging-in-Publication Data
Names: Roy, G. C. Datta, author.
Title: Implementing energy efficiency in industries / GC Datta Roy.
Description: First edition. | Boca Raton, FL : CRC Press, 2024. |
Includes bibliographical references and index.
Identifiers: LCCN 2024019440 (print) | LCCN 2024019441 (ebook) |
ISBN 9781032532608 (hardback) | ISBN 9781032542140 (paperback) |
ISBN 9781003415718 (ebook)
Subjects: LCSH: Energy consumption.
Classification: LCC HD9502.A2 R69 2024 (print) | LCC HD9502.A2 (ebook) |
DDC 333.79—dc23/eng/20240805
LC record available at https://lccn.loc.gov/2024019440
LC ebook record available at https://lccn.loc.gov/2024019441

ISBN: 9781032532608 (hbk)
ISBN: 9781032542140 (pbk)
ISBN: 9781003415718 (ebk)

DOI: 10.1201/9781003415718

Typeset in Times
by codeMantra

Dedication

Contents

Foreword

I have realized during my 30 plus years at Dhampur Bio Organics Ltd. (and the undivided company Dhampur Sugar Mills Ltd.), becoming energy efficient is not a one-off project but a culture or a mindset. We were a sugarcane-processing company, which now produces bio-fuels, renewable energy, and sugar.

I truly believe that 'becoming energy efficient' in a cost-effective manner is one activity that any manufacturing industry cannot ignore, and it is not a one-off project but a life-long endeavour. Driving this culture to become a leader in energy efficiency will result in a substantial cost and competitive advantage across various industries, and championing this can be the difference between a good company and a truly outstanding one.

I was just a couple of years in work when my father, who was then leading the company, introduced me to Dr. Roy, and it was the beginning of a long relationship wherein we have worked on a wide gamut of projects and discussed even a wider range of subjects. The work we have done not only includes energy efficiency projects, but we also worked from design to implementing what was at that time one of the most energy-efficient bagasse-based cogeneration projects in the world. Over the course of our multi-decade relationship, our discussions have spanned subjects ranging from energy-efficiency to grassroot-level training and now to business strategy and developing a culture of learning and training in an organization.

Given Dr. Roy's ability to explain key technical issues in a simple and basic manner coupled with his passion on this subject will, I'm sure make this book a must read across a wide segment in any organization who want to embark on this path.

Gautam Goel
Managing Director,
Dhampur Bio Organics Ltd.

Preface

This book is a product of learning captured over close to three decades of work as a field engineer and another over two decades as consultant in the domain of energy efficiency and conservation (EE&C) in energy-intensive industries. The idea of writing this book was conceived many years back while preparing the response to a comment from a CEO of a large multinational paper conglomerate: 'OK Dr Roy, we have gone through your proposal on performance contract. It looks OK, but I was expecting something more exciting to make it worthwhile for the Board to spend time on'.

We spent more than an hour discussing what can be done to make the proposal 'more exciting'. The conclusion was that we have to find out the minimum energy needed for their production process, and together we would have to work to achieve the same. During the course of the discussions, it also emerged that the industries in the MSME segments face challenge in implementing energy efficiency measures due to their inability to attract and retain good engineering talents. As such, we also have to assess the capacity of the organisation to undertake high energy efficiency measures and the need for capacity building to sustain the activities.

That day I realised that as consultants, our standard menu-driven approach for energy audit and development of energy efficiency project must be changed. We redesigned our business processes to address the twin questions – minimum energy approach for energy audit, capacity assessment of the organisation and training for enhancing the same.

This book has been designed to act as a guidebook for the implementation of energy efficiency measures in industries – detailing out every individual step from audit to implementation in a user-friendly manner, backed with case illustrations. These case studies have been collated from the autor's notes and learnings from implementation of different types of projects in different industry segments.

The basic principles have been explained in a simple way, not dwelling too much on theories, so that readers with knowledge of simple physics and spreadsheets will be able to understand most of the equations, calculation routines, and graphics.

This book has been written for engineers and EE&C professionals engaged in design, engineering, construction, operation, and performance verifications of EE&C projects in industries. The contents can also be used for training of technicians deployed for energy audit, execution of retrofit projects, and measurements and verification of project performance post-execution.

Acknowledgements

Writing a book based on diarised information from the professional career spanning almost six decades is a substantial undertaking, and the author's family played a crucial role in making it a reality. There were many frustrating moments while progressing the preparation of the manuscript due to software glitches. The author sincerely appreciates the efforts of his daughter Anindita in overcoming these software challenges, as well as the contributions of his son Anindya and daughter-in-law Garima, both professional designers in preparation of the drawings and enhancing the visual aspects of this book. The emotional support and motivation from the author's wife Jhumur were also instrumental in keeping the project on track. The author's perseverance in preparing the manuscript for over a year, despite moments of doubt, is a testament to their dedication. This book truly reflects a collaborative effort, with the entire family playing a vital role in its development.

About the Author

GC Datta Roy graduated with honours in electrical engineering (Hons) from Jadavpur University, India in 1965. He was awarded Ph.D. by BITS, Pilani, India, in 1996. He is also an alumnus of the Executive Programme of Darden School of Business, University of Virginia, USA, in 1990.

He had undergone intensive trainings in energy efficiency and conservations including in cogeneration with the then Swiss MNC Brown Boveri & Co in Switzerland in 1970 and energy efficiency and conservation with Asian Productivity Organisation (APO), Japan, in 1980.

He has over four decades of experience in the energy-intensive manufacturing sector followed by another two decades of consulting in the domains of energy efficiency and renewable energy.

As a CEO and later as an advisor of a leading ESCO, he has been involved in providing policy advisory services as well as implementation services for large number of energy efficiency and renewable energy projects in the industrial and commercial sectors in many countries, including Algeria, Bangladesh, Egypt, France, China, Ghana, India, Ivory Coast, Jordan, Kenya, Malawi, Mongolia, Nepal, Nigeria, Pakistan, Singapore, Spain, Sudan, the UAE, and the USA.

Contribution of Dr. Datta Roy in the energy efficiency domain has been globally recognised by conferring of several awards such as 'Recognition certificate' from IPMVP (Currently known as EVO), USA, 2007, 'Recognition certificate', Dubai Energy Forum, Government of UAE, 2010, 'Energy Professional Developer of the year' by India chapter of AEEE, USA in 2010, and as 'Sustainability leader of the year' by Parivartan Sustainability Outlook in 2012.

Widely travelled, he has published large number of articles in industry journals and magazines and made presentations in many conferences across the globe.

1 Introduction

ENERGY EFFICIENCY IMPERATIVES

Energy efficiency and conservation is crucial for businesses as it helps reduce energy consumption, lower costs, and decrease emissions, making operations more sustainable. Energy-efficient technologies often go hand in hand with process optimisation and automation. By upgrading equipment, improving energy management systems, and adopting energy-efficient practices, businesses can streamline their operations, reduce downtime, and enhance productivity. Increased productivity leads to higher output, improved quality, and ultimately, greater profitability. In addition to the direct benefits, energy efficiency in industries results in many indirect economic and social benefits.

The adoption of energy-efficient technologies creates job opportunities across various sectors. Manufacturing companies involved in producing energy-efficient equipment and components experience increased demand, leading to job growth in their industry. The pursuit of energy efficiency drives innovation and technological advancements. Businesses invest in research and development to develop more energy-efficient technologies and solutions. This not only leads to improved energy performance but also fosters economic growth through the development of new products, services, and markets.

Energy efficiency plays an important role in enhancing energy access, particularly in developing regions. Energy efficiency measures can help extend the reach of energy services to more people, including those in remote areas due to reduced demand from the industry. Energy efficiency also contributes to energy security by reducing dependence on imported energy sources.

The concept of energy efficiency as the "first fuel" highlights its importance in the energy transition. Before considering new energy sources or technologies, it is essential to maximise the efficient use of existing energy resources. The International Energy Agency (IEA) suggests that existing commercially proven technologies alone are sufficient to double global energy efficiency by 2,040. This shows that we have the means to achieve substantial energy savings using available tools and practices. However, it requires widespread implementation and adoption of energy efficiency measures across all sectors, particularly the industry.

The application of emerging technologies, such as digitalisation and artificial intelligence (AI), has the potential to further enhance energy efficiency gains through better conservation practices. These technologies enable advanced monitoring, optimisation, and automation of energy systems, leading to more precise control and better decision-making. Smart grids, energy management systems, and machine learning algorithms can optimise energy consumption patterns and identify opportunities for energy savings.

DOI: 10.1201/9781003415718-1

Industry accounts for over 30% of the global energy consumption.[1] Energy efficiency in industries will play a critical role in achieving the 'Net Zero' goal in conformity with the obligations of individual countries under various climate change treaties.

APPROACH FOR IMPLEMENTATION

Energy efficiency activities in industry follow both top-down (policy-driven) and bottom-up (engineering-driven) approaches. While policy-driven activities cover the sector as a whole, the individual industries adopt an engineering approach to identify and implement specific measures. Energy audits are universally used as the most effective tool to determine the extent to which energy consumption can be reduced by specific measures. Conventional energy audits focus on individual equipment efficiency and often overlook system-level investigations due to complexity and high costs. However, conducting energy audits at the system level can lead to significantly greater energy savings for industries. This requires a change in the approach to the energy audit process. Auditors, in consultation with the host industry, can set ambitious goals (such as doubling the efficiency gain by 2030) and conduct the audit driven by these goals. For this, we use higher-order analytical tools, such as waste heat analysis, process synthesis, exergy analysis, heat exchanger networking, digitalisation, and AI. The competency required for carrying out such analyses is quite high, and this may raise the costs of audit.

With digitalisation, it is now possible to develop in-house tools for field engineers, empowering them to perform advanced analyses. A step-by-step approach is typically followed, starting with an understanding of the processes, development of process flow diagrams, energy flow diagrams, and piping and instrumentation diagrams. Analysing the energy flow provides insights into the optimum energy requirements of the process, current usage, and the gap that can be bridged. Goals and targets are then set to bridge this gap. Basic knowledge of thermodynamics, electro-mechanical systems, and energy conversion processes is essential for developing this process.

The goal-setting process can also be top-down or bottom-up. Top-down approaches driven by regulations are prevalent in both advanced and developing countries. Examples include China's Top–1,000 and Top–10,000 enterprise programs, India's PAT (Perform, Achieve, and Trade) program, and the USA's SEP 50001. These programs set targets for reducing specific energy consumption in energy-intensive industries such as thermal power generation, cement, iron and steel, aluminium, chemicals, textiles, pulp and paper, and others. These targets are periodically revised based on past achievements and technological advancements.

SCENARIO

The efficiency scenario in industries is characterised by a few high-performance leaders, a few in the middle tier trying to catch up, and a significant number of laggards, who are unable to keep pace with the development due to various reasons such as finance, organisation capacity, and market distortion. The high-performing companies set benchmarks for energy efficiency and carbon intensity of their operations;

they take risks and invest heavily in state-of-the-art technologies. They are mostly the global players having access to both technologies and finances, while the companies in the middle and lower tiers are predominantly from emerging economies.

BRIDGING THE KNOWLEDGE GAP – BOOK OBJECTIVE

There is a considerable knowledge gap amongst engineers and technicians in these three categories of industrial groups. The industries in the third category and to some extent in the second category find it very difficult to attract and retain talents and build capacity to undertake energy efficiency improvement measures. The only option they have is to build capacity by providing adequate and relevant training to various staff involved in operation and maintenance (O&M).

The contents of the book have been developed to serve as a reference for professionals involved in energy management and sustainability efforts within industrial sectors, helping them make informed decisions and take practical steps toward greater energy efficiency. The book has been organised in a structured manner, with chapters and sections that follow a logical progression from fundamental concepts to implementation of projects and verification of results. This organisation facilitates a step-by-step learning process for the energy management professionals. Educational institutions and training programs can use the book as a foundation for developing energy efficiency training curricula. It can serve as a textbook for courses dedicated to energy efficiency in the industrial sector.

ENERGY EFFICIENCY AND CONSERVATIONS – BASIC PRINCIPLES

Fundamentals of energy efficiency and conservation are derived from the laws of thermodynamics. The first law, also called the law of conservation, states that energy can neither be created nor destroyed but it can be converted from one form to another. The second law, also called the law of entropy, governs the process of converting heat to work or vice versa.

Energy systems in most of the energy-intensive industries consist of utilities and process energy systems. Utilities consist of the supply side system such as the conversion of fuel energy into thermal energy and thermal energy into mechanical and electrical energy and the demand side system such as electro-mechanical devices. These devices include pumps, air handlers, compressors, material handling equipment, and more. Almost all energy-intensive industrial processes require the addition or removal of thermal energy. Examples include heating, cooling, evaporation, condensation, distillation, crystallisation, and many others.

The thermal and thermo-dynamic principles articulated under the laws of thermodynamics govern the conversion processes and efficiency of utilisation of the fuel heat for production processes. These principles and the processes have been explained in a simplistic manner in Chapter 2.

Chapter 3 provides a brief description of the processes and the energy flows for few of the energy-intensive industries. These industries have been chosen based on the designation of such energy-intensive industries by IEA, EU as well as countries such as the US, China, and India amongst many others. Various tools used for the analysis of the process energy systems and targeting energy savings potential have been described with illustrative examples. The information provided sets the context for energy audit and higher-order energy system analysis for implementation of different kinds of energy efficiency projects in industries overall and energy-intensive industries in particular.

ENERGY AUDIT AND IMPLEMENTATION OF PROJECTS

Energy audit is used for investigative analysis of the energy use in a facility, identification of wastages and measures for reduction of the waste. Different types of audits are practised depending on the needs of a particular industrial facility. The types of audits and their applications have been described in Chapter 4. Sufficient details have been provided on the 'Whats' and 'Hows' of the processes along with practical examples.

Energy audit is carried out by both internal and external teams trained for the job. In the case of external audit, better value is derived by proper scoping and documentation for the deliverables. The approach and methodology for the same have been described in the chapter. The outcome of the audit consists of a bank of feasible retrofit projects.

A retrofit project can be a simple one such as the replacement of an inefficient pump by a more efficient one or a very complex one such as a waste heat recovery project in an integrated steel plant. There would usually be an array of projects with varying degrees of complexity in between. Emerging computational technologies such as digitalisation and AI have opened floodgates for exploration and exploitation of new energy savings opportunities. The use of such technologies leads to additional energy savings through better management of information and operational control. Along with opportunities, we are also facing new challenges due to rapid obsolescence of such technologies. The industrialists particularly in the MSME (Micro, small and medium enterprises) segment now want faster payback from investment in energy efficiency for countering the impact of technology obsolescence and changes in the market dynamics.

Several implementation options are there for financing, engineering, procurement & construction (EPC) of projects. These include Turnkey EPC, BOOT (Build, own, operate & transfer), and ESCO (Energy services companies) performance contracts. Every organisation needs to develop its own implementation strategy depending upon the availability of finance, internal skill and competency and risk management capacity. For example, the simple project of pump replacement can be executed in-house while outsourcing is a better option for complex projects, such as waste heat recovery. The different implementation models have been discussed with illustrative case studies in Chapter 5.

MEASUREMENT AND VERIFICATION

How do we capture the energy savings as a result of implementation of a project? Even in a simple project such as the replacement of a pump, actual energy consumption can increase as a result of longer hours of operation post-implementation of a retrofit project. In such a case baseline energy consumption has to be determined afresh considering the longer hours of operation. It is important to develop a common understanding of the energy consumption baseline and adjustment principles. Protocols for measurement and verification (M&V) of different types of projects known as International Performance Measurement & Verification Protocol (IPMVP) are used for this. The M&V is performed in two steps. The first step is the preparation of an M&V plan, which becomes a key component of the EPC document and then carrying out the verification as per the plan. Chapter 6 provides details of the process as well as few illustrative examples of M&V of implemented projects in industries.

EMPLOYEE ENGAGEMENT

Employee engagement plays a pivotal role in driving energy efficiency initiatives within an organisation. When employees are actively involved and motivated to reduce energy consumption, it leads to significant cost savings, environmental benefits, and a more sustainable workplace. Commitment from the leadership demonstrated by their active participation in the energy management program sets the process of engagement. The tools for keeping the employees engaged are well known. These include setting clear targets and goals, sharing of information on programs and achievements, empowerment through training and development, continuous improvement through EnMS energy management systems, rewards and recognitions and above all fostering a culture of openness where everyone can freely share their ideas and feedback on what is working and what things need to improve.

Realising the critical role of employees in creating a sustainability-focused organisation, top companies across the globe are innovating their programs within the framework of the engagement model. A review of these models indicates that information and communication are the two common pillars on which to rest all such programs. Digitalisation has revolutionised the processes of generating information and sharing the same through various modes of communication platforms such as social networks. However, this is also creating some problems due to loss of private time and information overload. A prudent practice will be to design an information-sharing and communication protocol which will fully address the needs at different levels without causing an overload.

It is the author's view that the employee engagement process will be the main driver for implementation of the energy efficiency programs in the industries. Chapter 7 has, therefore, been made a part of this book to sensitise the readers on happenings on this front and how they can play their important roles in creating a sustainability-focused organisation.

NOTE

1 IEA, CO_2 emissions in 2022, 2023, www.iea.org/reports/co2-emissions-in-2022.

2 Fundamentals of Energy Efficiency and Conservations

INTRODUCTION

In this chapter, we delve into the scientific and thermodynamic principles that underpin efficiency and conservation in the context of energy systems. We also explore what is understood by some of the terms that are used to define energy efficiency, terms such as 'energy effectiveness', 'first law efficiency', 'second law efficiency', and 'energy productivity' commonly used in industries to better comprehend the technical and financial aspects of energy system performance.[1] While there are subtle distinctions between efficiency and conservation, both concepts share the overarching goal of enhancing the energy performance of a facility. However, it's worth noting that in many cases energy savings can be achieved through simple conservation practices not requiring any capital investment.

Efficiency refers to how well an energy system converts the input energy into useful output energy. It's often expressed as a percentage, where higher efficiency indicates less wasted energy and better overall performance. The concept of efficiency is governed by the First Law of Thermodynamics, which states that energy cannot be created or destroyed, only converted from one form to another. Therefore, improving efficiency means maximising the useful energy output while minimising energy losses.

On the other hand, conservation involves reducing the total energy consumption of a system without necessarily improving the efficiency of energy conversion. Conservation strategies aim at decreasing the overall energy demand by implementing good operating practices. Switching off a device when not required, regulating the water flow as per process need, and maintaining clean heat exchanger surfaces are some of the simple examples of conservation measures.

It's important to note that while efficiency improvements often require investments in technology and equipment upgrades, conservation measures can sometimes achieve significant energy savings without major capital expenditures.

Energy effectiveness is more of an economy term which means for what purpose we are using it, how energy is adding value in an overall economic system. If it is not adding value, then we are wasting the energy however efficient the conversion process may be.

An industrial energy system consists of a series of energy conversion processes that involve transforming fuel into heat and heat into electricity and mechanical work to drive various industrial processes. Each of these conversion processes presents an opportunity to reduce energy consumption through improvements in

DOI: 10.1201/9781003415718-2

first- and second-law efficiencies. By understanding the thermodynamic principles that govern these individual conversion processes, we can identify potential areas for improvement.

In response to the oil shock of 1973, Japanese industries had extensively deployed conservation measures to maintain their competitiveness. We will see from a case study how Japan was able to decouple GDP growth from energy consumption, setting examples for the rest of the world to follow. This case study demonstrates the practical implementation of energy conservation strategies and their positive impact on economic and energy-related factors.

ENERGY SAVINGS – EFFICIENCY AND CONSERVATION

As stated earlier, energy can be saved by both efficiency and conservation measures. We use electrical energy for cooking, lighting, cooling, and water pumping in our day-to-day life. How do we define the input and output? At home, we have mostly one meter capturing the electrical consumptions for all the appliances. Thus, it is not possible to measure the energy consumption for individual appliances, unless we provide meter for every appliance or individual circuit. Similarly, we still do not have practical tools to measure outputs from the various appliances. That leaves us with the question of how do we get quality information that will enable us to undertake goal-driven programmes on energy efficiency and conservation? Digitalisation is expected to provide answers to most of the performance-related questions, thereby improving our ability to deal with information-related challenges better.

Sometimes, efficiency gains do not result in reduced energy consumption. This happens when we lose attention to conservation, what is known as the 'rebound effect'. Let us take the example of a 1,200 mm sweep ceiling fan used mostly in buildings. Power requirement of a traditional fan is 70–90 W as against about 25–28 W for a BLDC (brushless direct current) fan for almost the same amount of air displacement. Using a BLDC fan instead of conventional fan will make the fan-based cooling system three times more efficient. However, knowing that the BLDC fan consumes less energy, we may not care to stop the fan when not required. Suppose our cooling requirement is for 8 hours but we keep the fan on for 24 hours, the entire efficiency gain would be lost. Installing a BLDC fan is an efficiency measure while switching off the fan when not required is the conservation measure.

There can be overlaps between efficiency measures and conservation measures, and in some cases, certain measures can be considered both, thermal insulation being a case in point. The terminology used to describe these measures may vary, but what matters most is that we are implementing them to improve energy performance. Whether we classify them as efficiency measures or conservation measures, the ultimate goal is to achieve energy savings and reduce environmental impact. Efficiency measures are often specific interventions or technical solutions aimed at optimising energy use in a targeted manner. On the other hand, conservation can be seen as a broader culture or mindset within an organisation. It involves creating awareness, instilling habits, and fostering a collective effort to conserve energy throughout all levels and departments. Conservation goes beyond technical interventions and permeates the entire organisational structure, involving participation from shop floor workers, managers, and employees at all levels. This holistic approach to conservation can lead to the discovery of significant energy-saving opportunities that might otherwise go unnoticed.

In summary, while some measures may straddle the line between efficiency and conservation, the key is to embrace a comprehensive approach that combines targeted efficiency measures with a culture of conservation. By doing so, organisations can maximise their energy-saving perfromance and achieve sustainable energy management.

The following case study is a simple example of conservation through such behavioural intervention in an industrial unit.

Case Study 2.1: 40% Energy Savings through Conservation

This case pertains to the curing and bagging section in a sugar mill. Sugar manufacturing process involves extraction of juice by crushing of sugar cane, clarification of juice followed by concentration, crystallisation and separation of sugar crystals from the massecuites in centrifuges. Centrifuges operation was being carried out in three stages. The first and second stages (called B & C) operated in continuous mode, while the third stage (called A) was in batch mode. The sugar factory had launched a campaign for reducing internal power consumption with a view to earn more revenue by exporting the saved power. One of the operators of the centrifuge station suggested metering of batch centrifuge station for saving energy. His logic was that it would be possible to reduce the cycle time by optimising the operation sequences; scraping of screens and washing and loading. We acted on his suggestion and installed energy meter for the batch line. Average baseline energy consumption was 160 kWh/T. The optimisation exercise was carried out along with monitoring of the separation efficiency and sugar quality. At the close of the exercise, the energy consumption was reduced to 100 kWh/T thus achieving 40% savings over baseline.

In industry, we would find such examples galore offering opportunities for substantial energy savings through conservation, illustrating a few such examples:

Loss of fuel heating value due to improper storage
Fugitive losses of fuel from the preparatory and conveying systems
Loss due to improper mixing of fuel and combustion air
Power potential loss in thermodynamic cycles due to operation at lower parameters
Loss due to leakages of steam
Loss of heat due to improper insulation
Lower heat recovery due to dirty heating surfaces
Improper distribution of liquid/vapour in process equipment
Inefficient operation of cooling towers

Differences between the terms efficiency and conservation have been highlighted to underscore the roles of technology and behaviour in reducing overall energy consumption. Going forward, we will be using the term energy efficiency or conservation to denote both efficiency and conservation.

THERMODYNAMICS AND ENERGY CONVERSION PROCESSES

Thermodynamics is the science that governs the processes of energy conversion and utilisation. Figure 2.1 captures the entire energy value chain from fuel to end use, as typically seen in the industry.

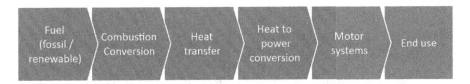

FIGURE 2.1   Industrial energy value chain.

Each of the element in this energy value chain offers opportunity for improving energy performance through energy efficiency and conservation measures.

Fuel

Industries use mainly fossil fuels; coal, oil and gas depending upon the availability and cost. Driven by the environmental consideration, biomass is also finding its niche as supplementary fuel even when such fuel may not be cost competitive. Off-take of other renewable energy resources such as solar and wind energy are also increasing for reducing the carbon content of the used energy.

Energy stored in fuel cannot be utilised directly. This is converted into usable energy by combustion. Thermal energy released by the combustion process is used for heating and power generation.

Fuel is physically characterised by the contents of volatile matters (VCM), fixed carbon (FC), moisture (M), ash (A) and the gross calorific value (GCV). These elements are determined by proximate analysis. Volatile matters consist of carbon (C), hydrogen (H) and oxygen (O) in gas form in the fuel. They get released during the early stage of combustion process and ignite. Higher volatile content therefore, reduces the ignition temperature. Higher levels of ash and moisture do not add to calorific value, thereby reducing the GCV. Further, some amount of heat gets consumed in evaporating the moisture in the fuel and moisture generated from combustion of hydrogen. The amount of heat available for productive utilisation is therefore, less than the GCV accounting for this lost heat in evaporation. The term net calorific value (NCV) is used to denote the amount of productive heat in the fuel.

FC is a measure of the amount of non-volatile carbon remaining, it represents the portion of the fuel that must be burnt in a solid state. Higher the amount of FC, higher will be calorific value. Different laboratory and online instruments are used for analysis of the fuel and calculation of the share of different components (physical as well as chemical).

Proximate analysis provides data on physical parameters while ultimate analysis is carried out to determine the share of individual chemical components including the caloric components. We calculate the GCV of fuel from the share of individual caloric elements and their calorific values. Table 2.1 shows calorific value (CV) of the individual caloric element in the fuel.

TABLE 2.1
Calorific Values of Components

Component	CV (kcal/kg)
Carbon	8,080
Hydrogen	34,500
Sulphur	2,240

We accordingly, calculate the GCV of the fuel from the share of individual caloric components (%) and applying the Dulong's formula as below.

$$GCV = \frac{1}{100}\left[8,080\ C + 34,500\left(H - \frac{O}{8}\right) + 2,240\ S\right] kcal/kg$$

GCV of solid fuel at the time of use can be lower than the value during purchase due to degradation during storage. Such degradation takes place due to spontaneous combustion and accumulation of water during rains. Storage bins are designed so that they are well ventilated with arrangement for drainage of rain water. It is also necessary to provide for adequate space between bins so that the handling vehicles can move freely. This helps in managing the inventory allowing for FIFO (First in first out) principle and reclamation of degraded fuel (Figure 2.2).

It is a good practice to carry out physical stock taking periodically. It is not uncommon to find that the physical stock is less than the book stock. Such things happen due to fuel degradation and actual operating efficiency being less than the reported efficiency. In addition to the benefit of early action on operating

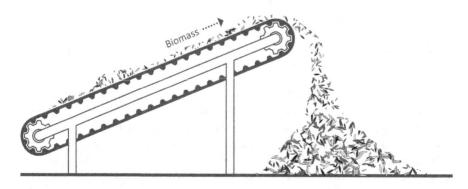

FIGURE 2.2   Reclamation of degraded biomass fuel.

inefficiencies, periodic monitoring helps in identification of hot spots due to spontaneous combustion.

Normally, we do not lose heating value of liquid or gaseous fuel during storage.
However, there are safety issues that require careful consideration.

It is best to design and operate the fuel storage and handling systems as per standard global or country specific protocols.

COMBUSTION

Combustion is the process of rapid oxidation of fuel with production of heat, or heat
and light. Complete combustion (complete oxidation) of a fuel is possible only in the
presence of an adequate supply of oxygen and intimate mixing of the caloric elements with oxygen. Oxygen required for combustion is supplied by the oxygen in the
fuel and supplied combustion air.

3Ts: temperature, turbulence and residence time are the pre-requisite for proper
combustion of the fuel (Figure 2.3).

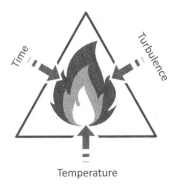

Temperature

FIGURE 2.3   3Ts of combustion.

Temperature above the ignition point is required for the fuel to ignite
Turbulence ensures proper mixing of fuel and air reducing the need for excess air
High residence time that is more time in the combustion zone ensures complete
  combustion

The quantity of air required for complete combustion of the fuel, also known as
stoichiometric requirement, is calculated taking into account the share of the caloric
component and applying the combustion equation.

The products of combustions are:

$$S + O_2 = SO_2 + Heat$$

$$2H_2 + O_2 = 2H_2O + Heat$$

$$C + O_2 = CO_2 + Heat$$

TABLE 2.2
Stoichiometric Air Requirement

Element	Composition %	Oxygen Required/Unit kg/kg	Actual Oxygen Required kg	Air Required kg
Carbon	45	2.67	1.20	5.22
Hydrogen	5	8	0.4	1.74
Sulphur	0.5	1	0.005	0.022
Oxygen	18	−1	−0.18	−0.78
Total			1.43	6.20

Taking into account the molecular weight of the individual components, 12 g of carbon combusts fully with 32 g of oxygen producing 44 g of carbon dioxide or 1 g of carbon requires 2.67 g (32/12) of oxygen for complete combustion. Stoichiometric oxygen requirement for other combustibles is calculated the same way. Actual amount of oxygen requirement is computed after netting for the oxygen present in the fuel. Air requirement is calculated taking the oxygen content in the air at 23% by weight. Table 2.2 shows the calculated requirement of oxygen and air for a particular fuel composition.

In case of gaseous fuel, the molecules are diffusing in their minute state. That makes them very susceptible to contact with the oxygen molecules. In case of liquids, the molecules are combined and therefore, not able to develop intimate contact. Liquid fuels have to be therefore, vaporised to individual molecules for developing the contact. Solid fuel needs to be disintegrated into fine particles for establishing similar contact.

Increasing the air improves combustion but this also increases the loss through the flue gas. The suppliers of boilers and furnaces specify the excess air requirement for the specified fuels. However, in the world of volatile fuel market we may not get the fuel of our choice. It is a good practice to carry out periodic efficiency tests for different fuels, assess the combustion losses and optimise the fuel air ratio to achieve the highest overall efficiency.

The loss due to unburnt carbon in ash due to incomplete combustion and presence of CO in flue gas due to partial combustion is calculated applying the following formulae.

$$L_1 = \frac{\text{Total ash/kg of fuel burnt} \times \text{GCV of ash}}{\text{GCV of fuel}}$$

L_1, loss due to unburnt carbon in ash.

Similarly, loss due to CO on account of incomplete combustion is calculated as follows.

$$L_2 = \frac{\%CO \times C}{\%CO + \%CO_2} \times \frac{5{,}744}{\text{GCV of fuel}} \times 100$$

L_2, loss due to CO; CO, volume of CO in flue gas leaving furnace; CO_2, volume of CO_2 in flue gas; C, carbon content kg/kg of fuel.

Part of the heat gets consumed in evaporation of fuel moisture as well as moisture (H_2O) due to combustion of hydrogen (H).

$$L_3 = \frac{M\left(584 + c_p\left(T_f - T_a\right)\right)}{GCV \text{ of fuel}} \times 100$$

M, kg moisture in fuel on 1 kg basis; c_p, specific heat of flue gas in kcal/kg°C; T_f, fuel gas exit temperature in °C; T_a, ambient temperature in °C; 584, latent heat corresponding to partial pressure of water vapour in kcal/kg.

To assess these heat losses, carbon, ash, hydrogen and moisture content in the fuel is determined by ultimate analysis. Carbon in ash is determined by loss of ignition test. CO is determined using flue gas analyser.

Various measures for reduction of these losses include improving fuel preparation, better distribution of primary and secondary air and optimisation of air-fuel ratio.

Part of the heat also gets lost in the flue gas depending upon the efficacy of the amount of heat recovered through the heat transfer system described as follows.

HEAT TRANSFER

Heat transfer is a thermal process of transferring heat from one medium to another medium when they are existing at different temperatures. This transfer takes place through radiation, conduction and convection. Figure 2.4 illustrates the mechanism of heat transfer in a simple system such as cooking of food using an electrical heater (heater can be replaced by a gas burner).

Radiation is the energy emitted by matter in the form of electromagnetic waves or photons which do not require a medium to travel. It is the fastest way of heat transfer as the waves are travelling at the speed of light.

Conduction takes place through medium; solid, liquid and gas. In solids, it is due to the combination of vibrations of the molecules in a lattice and the energy transport by free electrons. In gases and liquids, it is due to the collisions and diffusion of the molecules during their random motion.

Convection is the mode of energy transfer between a solid surface and the adjacent liquid or gas that is in motion. It involves the combined effects of conduction and fluid motion. Convection heat transfer depends on the fluid properties dynamic viscosity (μ), density (ρ), and fluid velocity (v).

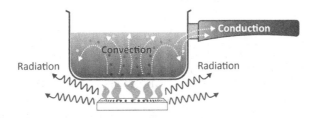

FIGURE 2.4 Heat transfer for water heating.

Heat is transferred directly in metal furnaces into the materials under process while indirect methods involving heat exchangers are used for generating steam for producing power and process heating. Heat transfer in the former case is mainly through radiation while it is through convection in other application.

Significant amount of heat is wasted in the direct process as the exit flue gas temperature is high. Part of the waste heat is recovered in preheating of the metal and combustion air. Even after that enough heat remains that can be commercially exploited using different type of waste heat recovery technologies.

Heat exchanger as the name implies is a mechanical device that transfers heat from one medium to the other without mixing the mediums. A cooking pan illustrated in Figure 2.4 is a heat exchanger transferring the heat of gas fuel to water. The outer body of the pan is getting heated up as a result of radiation waves, convective flue gas movement around it and conduction from the hot gases touching the outer surface. The inner body of the pan is heated up due to conduction from the outer body. Water molecules in touch with the inner surface heats up through conduction from the hot surface. This sets up the convection process in the water body due to density difference completing the transfer of fuel heat to the water. Many different types of heat exchangers are used in industries depending upon the fluid properties, available temperature differentials and techno-economics considerations. Shell and tube followed by plate exchangers are the most commonly used heat exchanger in industries (Figure 2.5).

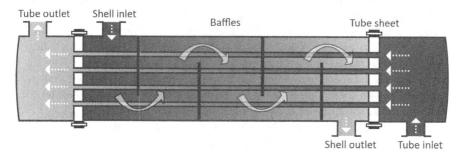

FIGURE 2.5   Shell and tube heat exchanger.

The heat transfer between the shell and tube side takes place through conduction and convection. In a perfectly insulated system, the rate of heat transfer (Q) from the shell side to the tube side or vice versa can be calculated from the properties of the fluid, fluid flow and the temperature differential between inlet and outlet. The simple equation for this is:

$$Q = m \, C \, \Delta T$$

where Q, heat transfer rate in kcal/h; m, fluid flow rate in kg/h; Cp, average specific heat in kcal/kg/°C; ΔT, mean temperature difference between inlet and outlet in °C.

The specific heat of a fluid has co-relationship with temperature. Let us take the case of water. It follows a hyperbolic trajectory with the lowest value of about 4.17 J/g°C at temperature of 40°C and the highest value is 4.21 at 95°C (Figure 2.6)

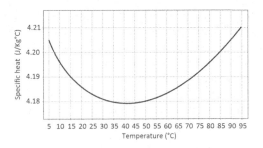

FIGURE 2.6   Specific heat of water.

The range of variation is not very high. As such, we can take a mean value that is about 4.19 J/g°C for evaluating efficacy of an operating water to water heat exchanger in an operating plant. Similar approach can be followed for most of the fluids except for the highly viscous fluid where the viscosity changes considerably with change in temperature. In such cases, it is necessary to determine the exact values over the temperature range and apply the required correction factor.

The heat rate equation above is valid when the phase of the fluid remains the same during the process of heat transfer. In case of phase change, Q is calculated from the rate of flow and the latent heat of vaporisation or condensation. The equation for this is:

$$Q = mL$$

where Q, Heat transfer rate in kcal/h; m, fluid flow rate in kg/h; L, latent heat in kcal/kg.

Having known the amount of heat to be transferred, we calculate the heat transfer area applying the following formula.

$$Q = h\,A\,\Delta T$$

where Q, total heat transferred in kcal/h; h, overall heat transfer coefficient in kcal/m²h °C; A, heat transfer area in m²; ΔT, mean temperature difference between the hot and cold stream in °C.

Heat transfer coefficient depends on both the thermal properties of a medium and the hydrodynamic characteristics of its flow. Overall heat transfer coefficient is a function of heat transfer coefficients of either side of the heat exchanger governed by the following equation.

$$1/h = 1/h_i + 1/h_o$$

where h_i, Heat transfer coefficient inside; h_o, Heat transfer coefficient outside.

Higher the value h, lesser is the requirement of heat transfer area for transfer of same amount of heat. Higher h value is achieved by increasing the velocity of fluid and turbulence. Replacing shell and tube heat exchanger by plate heat exchanger is an option for this. Some amount of extra power is required for pumping due to higher pressure drop with increased fuel flow. This partially offsets the gain from higher heat recovery. We have to therefore, carry out a cost benefit analysis for decision on substitution.

Quality of steam plays an important role in maintaining heat transfer efficiency. The steam to fluid heat exchangers is designed to operate on saturated steam.

Steam condenses during the process of heat transfer resulting in high heat transfer rate. In case we supply superheated steam, part of the heating surface would be utilised in converting superheated steam to condensing steam. This means lower productivity and efficiency. Similarly, some amount of heat of steam gets consumed in vaporisation of moisture in case of wet steam.

Over a period of time heat exchanger performance deteriorates due to scaling or fouling. The scaling occurs mostly from dissolved inorganic salt of calcium and magnesium while slagging and fouling occurs from deposition of solids from the dust laden flue gases from furnaces (Figure 2.7).

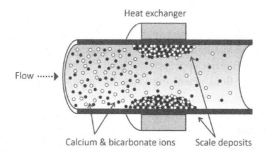

FIGURE 2.7   Scaling and fouling of heat exchangers.

Both preventive and curative steps are taken for keeping the heat transfer areas clean. Regular monitoring is done to assess the performance and take measures for cleaning, descaling as required.

The following example illustrates the example of a boiler, methodology for monitoring and diagnosis of non-performance of the various sections.

A steam boiler consists of a furnace for generating heat from combustion of fuel and a series of heat exchangers for transfer of heat from flue gas to water and steam (Figure 2.8).

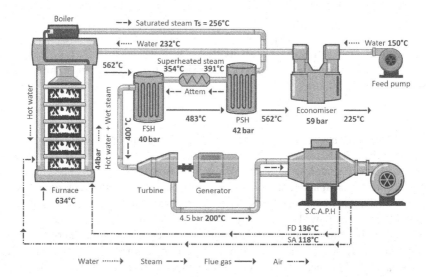

FIGURE 2.8   Boiler as a series of heat exchangers.

This boiler comprises of six heat exchangers, the furnace, final superheater (FSH), attemperator, primary superheater (PSH), economiser and steam coil air preheater (SCAPH). Bulk of the heat is transferred by radiation in the furnace zone and by convection in other zones. Conduction from the outer surface of the tubes to the inner surface is common for all the cases. These heat exchangers are designed for gradual transfer of heat of the flue gas to water and steam in different sections of the boiler so as to achieve the desired steam output at lowest specific fuel consumption. The flue gas is exhausted into the atmosphere after extracting the maximum amount of heat in these exchangers. The heat loss with the exit flue gas is calculated applying the following formula.

$$L_4 = \frac{m \; c_p \left(T_f - T_a\right)}{\text{GCV of fuel}} \times 100$$

where m, mass of dry flue gas in kg/kg of fuel; c_p, Specific heat of flue gas in kcal/kg °C; T_f, flue gas temperature in °C; T_a, ambient temperature in °C.

Loss with the flue gas is directly proportional to the differential between ambient and flue gas exit temperature. Performance deficiency in any of the heat exchanger section is likely to raise the exit temperature and therefore, heat loss. This loss can be minimised by monitoring and controlling the temperature drops across each section as shown in Figure 2.9.

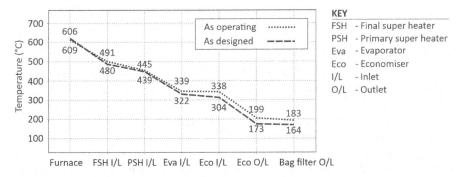

FIGURE 2.9 Monitoring of heat exchangers.

We are losing extra heat as the flue gas exit temperature post last recovery device (economiser) is 198.9°C against design value at 172.6°C, higher by 16.3°C. The temperature drop in the economiser section is only 2.4°C against design value 17.8°C. This can be due to either scaling inside the tubes or fouling outside. Scaling happens due to poor quality of feed water. Fouling in the economiser section consists mainly of loose dust that can be removed by soot blowing.

Similar methodology is adopted for monitoring of various process heat exchangers. In addition to energy savings, good performance of heat exchangers positively impacts process productivity too.

THERMAL INSULATION

Exposed hot surfaces lose heat to the atmosphere following the three principles of heat transfer that is radiation, conduction and convection. Loss through radiation is negligible at low temperature. It rises exponentially with increased temperature. Loss by radiation from a steel surface at about 200°C is almost six times more than from the surface at about 35°C (Oliver Lyle[2]) considering surrounding temperature at 20°C. Loss due to convection rises proportionately with temperature due to more rapid displacement of the hot air. Table 2.3 shows the calculated loss of heat from a bare steel pipe carrying hot fluid at different temperatures.

TABLE 2.3
Loss of Heat From a Bare Pipe

Fluid Temperature	Bare Steel Pipe Diameter (mm)						
	25	50	100	150	200	250	300
°C	Heat Loss (kcal/m/h), Ambient Still Air at 20°C						
100	98	175	330	477	625	773	921
150	201	358	677	978	1,281	1,584	1,887
200	327	582	1,099	1,588	2,079	2,571	3,063
250	479	853	1,611	2,328	3,048	3,768	4,489
300	658	1,171	2,212	3,179	4,186	5,175	6,166
350	863	1,537	2,903	4,196	5,494	6,792	8,093
400	1,116	1,987	3,753	5,424	7,101	8,779	10,460
450	1,378	2,452	4,632	6,695	8,766	10,837	12,913
500	1,666	2,966	5,602	8,096	10,600	13,106	15,616

Actual loss is usually more due to higher air velocity. Similarly actual loss is more in case of lower ambient temperature and less in case of higher temperature. The hot surfaces such as furnaces and boiler walls, steam pipelines and cold surfaces such as cold rooms are insulated to restrict this unwanted transfer of heat to the surrounding. In case of furnaces operating on batch mode, some quantity of heat gets absorbed in the refractory material during the heating cycle. This heat gets released during the cooling cycle leading to loss. The key properties of insulation materials are therefore, low thermal conductivity that prevents heat transfer and low heat capacity that minimises stored heat in batch processes.

Apart from the thermal conductivity and heat capacity, other important properties are

Working temperature range
Physical and chemical strength
Durability
Fire resistance
Water vapour permeability
Adaptability to the construction site and finally
The cost per unit volume

Ceramic fibre and mineral wool are the most commonly used insulations for hot surfaces.

TABLE 2.4

Insulation Impact on Heat Loss

Fluid Temperature °C	Thickness mm Loss (kcal/m/h)	Bare Steel Pipe Diameter (mm)						
		25	50	100	150	200	250	300
		Thickness of Insulation (mm) & Heat Loss (kcal/m/h)						
100	Thickness	25	30	40	40	50	50	50
	Heat loss	18	24	32	42	45	53	62
150	Thickness	40	50	50	65	75	75	75
	Heat loss	25	31	48	53	58	69	80
200	Thickness	50	50	75	75	85	85	85
	Heat loss	31	47	56	72	81	95	109
250	Thickness	50	65	85	100	100	100	100
	Heat loss	47	56	71	82	99	116	133
300	Thickness	65	75	100	100	120	120	120
	Heat loss	54	68	84	107	114	133	152

Heat loss in uninsulated pipes is more with higher diameter and with increased temperature. We have to therefore, increase the thickness of insulation with increase in diameter as well as temperature. Table 2.4 shows the insulation thickness for different conditions and their impact of heat loss (Authors note from Japanese standard 1980).

How thick should be the insulation? We can choose the thickness as per the values in Table 2.4. However, the situational variables such as the average annual ambient temperature, operating hours, energy price and cost of insulation have their impact on techno-economics, hence the final choice.

It is a good practice to carry out periodic temperature and heat loss surveys of outer surfaces of furnaces, boilers, pipelines and assess the losses. Such survey reveals the need for replacement, repair and augmentation of insulation based on findings from the survey and cost benefit analysis.

HEAT TO POWER – THERMODYNAMIC CYCLES

Heat from the combustion of fuel is converted to work deploying an appropriate thermodynamic cycle. A thermodynamic cycle consists of sequential processes through changes in pressure, temperature and phase of a working fluid. In the process, the entire amount of the heat cannot be converted to work, limited by the temperature of the cold sink. Rankine cycle followed by Brayton cycles for generation of electrical power and cogeneration of power and steam are extensively used in industry. Vapour compression and absorption refrigeration cycles are the other two cycles that are commonly used for cooling in industry, buildings and transportation.

In case of a steam turbine operating on Rankine cycle, superheated steam from boiler is expanded into a turbine producing the mechanical work. Part of the heat is converted to work and the balance is exhausted into a condenser wherein the vapour is condensed giving up the latent heat to the circulating cooling water. The hot water is cooled in a cooling tower and recirculated. The condensate is recirculated as feed water to the boiler after preheating for deaeration and improving cycle efficiency. The efficiency of a thermodynamic cycle is governed by the principle that entire energy in the working fluid cannot be converted to work. Part of the energy has to be dissipated as waste. We can accordingly write the following formula.

$$\eta th = \frac{W}{QH} = \frac{QH-QC}{QH} = 1 - \frac{QC}{QH}$$

where ηth, the thermodynamic efficiency; W, work done; QH, heat input; QC, unutilised heat discharged into the cold sink.

Efficiency is increased by increasing the input heat or reducing the unutilised heat. This requires increasing the temperature at the inlet or decreasing the temperature of the sink. We increase the input heat by increasing the inlet temperature. We do not have much control over the sink temperature as atmosphere is the final sink.

In case of Brayton cycle, highest temperature corresponds to the combustion temperature for the fuel. However, in actual practice, this gets limited by the safe temperature limit for the material in the hot section of the turbine. The temperature is maintained within safe limit by regulating the flow of the combustion air.

In case of Rankine cycle, higher efficiency is achieved by increasing the temperature of superheated steam from the boiler. However, with increase of the temperature, the entropy of the system increases thereby reducing the work potential. Simultaneously therefore, pressure is increased to contain the rise of entropy. Figure 2.10 show process flow and the thermodynamic cycle diagrams for the Rankine cycle.

Steam is generated in the boiler utilising the fuel heat (Q_H) and fed into the steam turbine wherein part of the heat gets converted to work (W). The remaining heat (Q_L) is exhausted into the condenser. The condenser is a shell and tube heat exchanger where cooling water flows through the tubes condensing the vapour exhausted from the turbine into the shell. The hot cooling water is cooled in a cooling tower (not shown in the diagram) and recirculated back in the condenser.

The heat power cycle diagram (Figure 2.11) shows the same process from the perspective of thermodynamics.

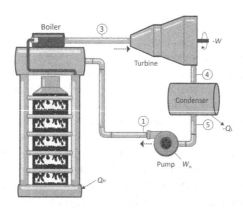

FIGURE 2.10   Process flow diagram-Rankine cycle.

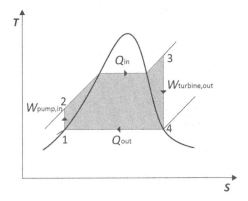

FIGURE 2.11 Rankine cycle diagram.

This diagram shows the temperature (Y-axis) and entropy (X-axis) relationship of the water-steam Rankine cycle. Process starts with pumping of hot feed water into the boiler (process 1–2), where it gets pre-heated to the saturation point following the liquid line of the saturation curve. The saturation temperature for the water is directly correlated to the pressure as is seen from the cycle diagram. Addition of further heat vaporises the water until the entire quantity of the water converts into vapour. Heat needed for converting saturated water to saturated vapour is called the latent heat of evaporation. This part of the process is represented by the heat addition in the economiser, furnace and the convection bank of the boiler. The vapour is further heated in the superheater section to reach the desired steam temperature for injection into the turbine. The lines 2–3 represents this entire transformation process.

Steam is expanded in the turbine represented by the line 3–4 delivering the mechanical power. This mechanical power is converted to electrical power by coupling a generator to the turbine. Post expansion in the turbine, steam loses the superheat getting back to the vapour phase represented by point 4 in the diagram. The vapour is condensed losing the latent heat in the condenser represented by point 1.

The area bounded by the line 1–2–3–4–1 represents the amount of heat converted to mechanical work. The larger the area, more is the work output. As explained earlier, the lower temperature that is temperature of point 1 depends upon the ambient temperature, hence we do not have control over this. Therefore, we have to increase the inlet temperature so that the upper line moves further upward thereby increasing the area. The pressure is increased so that we do not lose the work potential due to increase of entropy.

In this process, maximum heat is lost as the latent heat during condensation represented by the area bounded by line 4-1 and X-axis. We can increase the cycle efficiency by reducing the need for condensation. This is achieved in several ways:

Extracting and bleeding some amount of steam from the turbine before the exhaust stage for preheating of feed water. Both sensible and latent heat of the extracted/bled steam is utilised. This is called regenerative system.

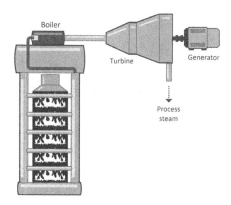

FIGURE 2.12   Ideal cogeneration system.

Extracting steam after partial expansion and reheat the same before admission
into lower pressure cylinder for further expansion up to the exhaust point.
This is called as the reheat cycle

Extracting and bleeding some amount of steam from the turbine for process
heating in industries. This is called cogeneration. In a perfectly balanced
case, entire quantity of the exhaust steam from the turbine is used in process
eliminating the need for condenser as illustrated in Figure 2.12.

Energy-intensive industries such as fertiliser, chemical, paper and pulp, petrochemicals, textile, sugar, edible oil, food processing require large quantity of steam. Many
amongst these industries buy power from the grid and meet steam demand from low
pressure boilers. These industries are potential candidates for cogeneration.

Metal industries use different types of furnaces. The temperature of exit gas from
these furnaces are high wasting large amount of heat. We can utilise this waste heat
for power generation in Rankine cycle or steam generation for process heat.

INDUSTRIAL MOTOR SYSTEMS

Motor system consists of electrical motors driving mechanical equipment such as
pumps, fans, compressors, conveyors used in industries. According to IEA, motor
systems account for more than 40% of the electrical energy consumed globally[3] (2011)
Improving efficiency of the motor system is less complex compared to thermal systems. Motor systems mostly operate as utility outside of core processes. It is easier to
measure the efficiency performance of a motor system at both the system and equipment level. Similarly, improving efficiency through retrofit measures is also simpler.
Considering the universality and replicability of the motor systems, quite a few global
alliances are working on continuous improvement of the motor systems. Examples
are global motor system efficiency initiative, US motor system efficiency initiative.[4]

Efficiency improvement measures include use of high efficiency motors and
equipment, reduce the losses due to friction and operating the system optimally. Let
us take the following example of a water pumping system for developing a clear
understanding of the energy usage and efficiency (Figure 2.13).

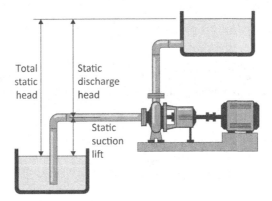

FIGURE 2.13 A pumping system.

The pump is lifting water from an underground tank and discharging into an overhead tank. A throttle valve is provided at the delivery end for regulating the flow. The pump has to generate a head equal to the total static head and the loss of head in the system on account of flow of water, called the dynamic head.

The power requirement and the efficiencies are calculated by measuring the heads (Pressure gauges at suction and discharge), water flow (In situ flow meter or portable meter such as ultrasonic flow meter) and power (In situ energy meter or portable power analyser) and applying the following formulae:

$$P_h = Q \times H_p \times \rho \times g/1000$$

where P_h, hydraulic power in kW; Q, flow of fluid in m³/s; H_p, total head in m=delivery head – suction head; ρ, density of fluid in kg/m³; g, acceleration due to gravity in m/s².
Efficiencies of the system and the pump are:

$$\eta_s = P_h/P_t$$

where η_s, system efficiency; P_h, Hydraulic power in kW calculated above; P_t, Input power at the motor terminal in kW as per meter.
The efficiency of the pump is:

$$\eta_p = \eta_s/\eta_m$$

where η_p, efficiency of pump; η_m, efficiency of motor (generally taken as per name plate data due to practical difficulty in measuring shaft power).

The system efficiency is high when each of the individual efficiency is high. High efficiency motor and pump is a choice exercised at the time of procurement. The efficiency of the rest of the system depends upon the design of the system and operational point, match between the pump and system characteristics.

The pump characteristics follow the affinity law relating pump flow (Q), head (H_p) and power (P_h) to the speed (N) as follows.

$Q \, \alpha \, N$
$H_p \, \alpha \, N^2$
$P \, \alpha \, N^3$

The pressure generated by the pump follows a drooping curve as shown in Figure 2.14. As the flow increases, the dynamic head increases due to higher pressure drop.

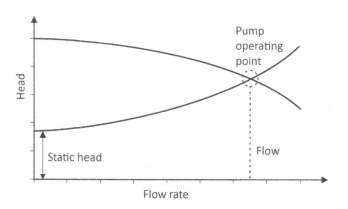

FIGURE 2.14   Pump and system characteristics.

In case of mismatch between the characteristics, efficiency is impacted by the method used for aligning the operating points as explained in Figure 2.15.

The actual flow requirement has reduced from flow 1 to flow 2 thereby reducing the system head requirement. At the required flow, the pump head increases following the characteristics curve corresponding to the flow 2. We can meet this requirement by taking either of the following measures:

Absorb the extra head in a throttle valve

Use variable speed drive for reducing the speed of the pump

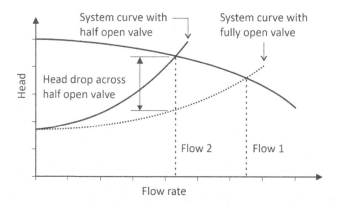

FIGURE 2.15   Variable speed drive for efficiency.

In case of throttling, the system becomes inefficient. The pump is still generating the extra head following its characteristic curve. The power required for this is wasted by dropping this head across the throttle valve increasing entropy of the fluid leaving the throttle valve. We can avoid this loss by reducing the speed of the pump to match the system point. Use of variable drive for energy conservation is now a common practice in industry.

Same principles apply for various other motor systems for improving the overall efficiency.

We have carried out over 500 energy audits of motor systems in various types of industries spanning the globe. The level of efficiencies has been found as low as 35% in some cases to as high as 84% in few throwing up huge energy savings opportunities. Intervention options are many as we would see from the case studies in the chapters to follow.

CONCEPTS TO APPLICATION – OVERCOMING THE BARRIERS

The imperatives of adopting efficiency and conservation are well known. Yet the uptake of investment in efficiency measures has not been commensurate with the potential and need. Historically, volatility of the energy market and cost of energy have remained the primary drivers for energy efficiency in industry. The environmental impact of energy consumption and global warming have recently raised the profile of energy efficiency in all sectors including industry.

Even then we face many barriers. There are market barriers such as energy pricing, subsidies; organisational barriers such as investment priorities and energy management structure, and behavioural barriers such as risk perception and discomfort with change. Three common refrains against allocation of resources for energy efficiency in industries are well known.

'Other priorities such as production do not leave much time for the senior management to look at efficiency'
'Plant engineers are uncomfortable with changes, high risk perception'
'Low skill competency of workers to operate and maintain high efficiency devices'

These barriers prevail due to lack of actionable information on benefits derived from efficiency for the organisation as well as for the individuals at different levels-CEO, plant engineers and the workers. Obviously, information need is different for different levels and changes over time. In today's time environmental considerations may be the main driver for CEO, for the engineers it would be measurability of energy savings while for workers ease of operation and maintenance.

Energy audit as a tool for identification of inefficiencies in energy use is a common practice and has been so from the mid-60s. However, it is the oil shock of 1973 that really triggered the first orchestrated movement for oil conservation worldwide. This became one of the key strategic responses for the oil importing countries towards energy security. Energy audit along with many other regulatory measures were adopted for implementing energy efficiency in all sectors of economy, even more so in industry.

Case Study 2.2: Decoupling Energy Consumption from Economic Development

Japan as a pioneer demonstrated that it was possible to decouple economic growth from energy consumption as shown in Figure 2.16 (Author's note from training report with Asia Productivity Organisation, Tokyo 1980).

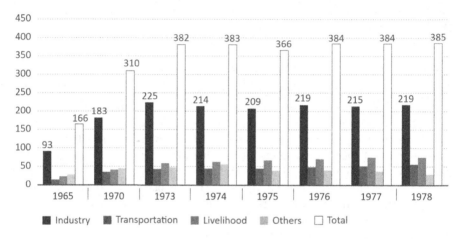

FIGURE 2.16   Japan's response to oil shock 1973.

Barring a marginal drop in 1975, Japan maintained GDP growth of around 5% without increasing energy consumption. This was achieved by actions on many fronts. Initial thrust was on conservation for achieving quick results. This was followed by large investment on technology for increased energy efficiency and then structural change moving away from natural resources to knowledge economy.

Energy cost is usually a more important driver for industry. Energy cost ranges from 15% to 40% of the total cost of manufacturing in energy-intensive industries such as steel, cement, aluminium, paper and pulps, and chemicals and petrochemicals. In a recent survey carried out in the industrial sector in the OECD countries, it has been found that cost control is over five times more likely to be a driver of energy efficiency than compliance with government regulations.[1]

Reduction of GHG emission is currently the most important driver at the levels of policy makers and CEOs. It is estimated that energy production and consumption account for over 66% of global carbon dioxide emission. Industry accounts for 24% of the total emission. IEA estimates (2022) that Industry's direct emission has to decline by nearly a quarter to 2030 in order to remain on track with the net zero scenario. This would mean reduction of emission by about 3% per year on average.[4] Drivers (energy security, cost reduction and GHG mitigation) are there. Cost effective technologies are also there. Even then the level of investment in such technologies are still not commensurate with the potential.

IEA (Energy Efficiency 2023) states that the greatest efficiency gains are achieved in all sectors with a package of policies that combines three main mechanisms: regulation, information, and incentives[5] for overcoming the barriers. While, policy makers deal with regulation and incentives, information remains the most important driver at the level of individual industries. Best results are achieved when information system is designed to support accountability of energy performance at each and every level of an industrial organisation. For the MSME segment too, information is one of the most critical barrier.[6]

The following case study shows a redesigned energy information system that facilitated investment decision on energy efficiency in a cement plant.

Case Study 2.3: Overcoming Information Barrier

A cement unit had carried out energy audit by engaging a reputed ESCO. The audit identified opportunities for savings of both thermal and electrical energy. It projected 12% reduction in the overall specific energy consumption by undertaking several short-, medium- and long-term measures. The identified measures were listed as individual projects with techno-economic analysis. Financial analyses were carried out adopting the investment analysis spreadsheet tool used by the host company. The findings were shared in a presentation meeting chaired by the head of the manufacturing operation. He raised the following questions:

The identified energy efficiency measures cut across various sections of the operations. What would be the methodology for monitoring the savings on a regular basis other than carrying out anther audit post implementation of the efficiency projects?

Further, even if the implemented projects perform as per design, the overall energy consumption may show increasing trend as a result of poor functioning of some other sections. In this case, how would we establish accountability? Are there any implementations issues such as production loss?

These are very valid questions. In addition to technical and financial issues, these questions address a much broader issues of information, organisation and management. It is possible to harmonise the technical and organisational issues by developing high quality energy information management system. Such information system can monitor both technical and management performance. The energy auditor should be able to understand the production management organisation and dovetail the energy information system to fit the production organisation. The process starts with developing performance indices (PI) for each of the individual manufacturing sections as illustrated in Figure 2.17.

The production manager is made responsible for the overall performance, his PIs are kWh/T Cement, kWh/T clinker. The PIs are then drilled down for individual manufacturing sections and further down to the next level. The impact of energy efficiency projects on PIs are calculated. The PIs are then revised for the different levels

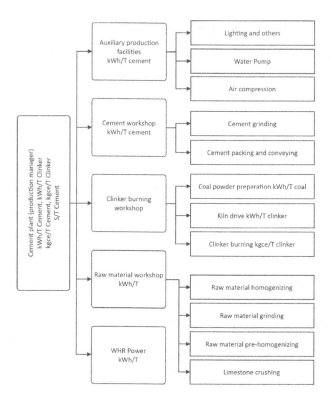

FIGURE 2.17   Performance indices.

bottom-up to the level of production manager. The company decided to implement a programme on 'energy information system' before undertaking implementation of investment projects.

Actionable information is the most important driver of energy efficiency and lack of it the most critical barrier. Digitalisation of the information system enables breaking of the information barrier. All levels of employees get access to the same information. Operators are able to make real time intervention through distributed control system (DCS) and other types of data acquisition system. The digitalisation will play the most critical role in removing information barrier and fostering a culture of energy conservation. A brief review of the role of digitalisation in improving energy efficiency and productivity follows.

ENERGY EFFICIENCY AND DIGITALISATION

What is digitalisation? In its simplest form, digitalisation means adaptation of digital devices for managing information system, technical processes and business transactions. Basic architecture of a digital system rests on the three pillars as shown in Figure 2.18.

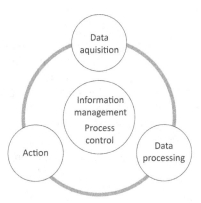

FIGURE 2.18   Digitalisation architecture.

Digitalisation involves the use of sensors, smart devices, automation, AI, and advanced data analytics to optimise processes and provide real time information such as trends for monitoring and control of the process variables. These systems use real time data to monitor and optimise energy consumptions. IoT (Internet of Things) devices and sensors can collect data on energy consumption patterns in industrial processes, analyse the same to identify inefficiencies, optimise operations, and make informed decisions to reduce energy waste. Digital technologies allow for virtualisation and remote monitoring of various processes and control of the same based on mobile applications.

The history of digitalisation spans over five decades. It has witnessed significant advancements in various technologies, leading to improvements in energy productivity. The beginning was made in early 70s with introduction of electronic programmable logic controllers (PLCs). The introduction of PLCs revolutionised industrial process control by replacing complex relay-based systems with digital logic controllers. PLCs allowed for flexible automation and easier reprogramming, enhancing efficiency and productivity in manufacturing processes. Simultaneously, Data Acquisition Systems enabled the collection and analysis of data from industrial processes, providing valuable insights for optimisation. DCS was the next important development that allowed integrating multiple control systems on a centralised platform. This enabled flexibility in design and alteration of the control algorithm based on gained experiences. DCS was the technology that revolutionised the empowerment of the process operators to optimise the systems for achieving better energy productivity.

Automation technologies, including robotic systems, gained prominence during mid-80s. Industries adopted automated assembly lines, material handling systems, and robotic arms, enhancing productivity and reducing manual labour. This allowed sustaining the efficiency levels at peak with reduction of human interface.

Artificial Intelligence (AI), Machine Learning (ML), 3D Printing (2000s–present) technologies brought intelligent automation and decision-making capabilities by the system itself without waiting for human interface. Each advancement of digital technology has made profound impact in improving energy efficiency and hence reduced GHG emission. Even then, it looks like that we have just made the beginning of harvesting the power of digital technologies, setting in process the industrial revolution version 4.

CONCLUSION

This chapter provides a comprehensive overview of the basic thermodynamic principles related to energy efficiency and conservation. It offers illustrated examples to help readers understand these principles more effectively.

The chapter explores the energy value chain in the industrial sector in fair amount of details. Each of the elements of the value chain offers opportunity for reduction of energy consumption through efficiency and conservation measures thereby improving the overall efficiency and energy productivity.

Key drivers of and barriers to the widespread adoption of energy efficiency and conservation measures have been highlighted backed by two case studies. One of the main contentions put forth by the author is that good quality information will be the key drivers for efficiency and digitalisation will play a crucial role in this. The chapter provides a brief overview of how digitalisation will facilitate the energy efficiency and conservation process. The discussions in this chapter sets the stage for subsequent chapters that delve into topics such as energy-intensive industries, energy audits, and the implementation of energy efficiency projects in industries.

NOTES

1 S. David Hu, *Handbook of Industrial Energy Conservation* (Van Nostrand Reinhold Company, 1983).
2 Oliver Lyle, *Efficient Use of Steam*, 1947 (Her Majesty Stationery Office).
3 Paul Wade and Conrad U Brunner, IEA, Energy Efficiency Policy Opportunities for Electrical Motor Driven Systems
4 IEA, Industry analysis, https://www.iea.org/reports/industry.
5 IEA, Energy efficiency analysis, https://www.iea.org/reports/energy-efficiency.
6 World Bank, India - Financing energy efficiency at MSMEs project (English). (Washington, DC: World Bank Group, 2013). http://documents.worldbank.org/curated/en/561461577729300064/India-Financing-Energy-Efficiency-at-MSMEs-Project.

3 Energy-Intensive Industries

INTRODUCTION

The First industrial revolution brought about significant economic growth and advancements through wide-scale industrialisation of the economy. The second industrial revolution, which occurred from the mid-nineteenth century until the beginning of World War I, built upon the foundations of the first industrial revolution and brought about significant advancements in technology and industrial processes. This period witnessed the widespread adoption and expansion of existing technologies, as well as the invention of new ones, leading to further economic growth and the globalisation of the economic system.

It is interesting to see that there was a close co-relationship between industrialisation and energy consumption. Energy was the prime mover driving the industrialisation. Large amount of energy was required for operating the metal extraction processes, transportation of both raw materials and finished products. The development of the railroad network was a major breakthrough, as it facilitated faster transportation of goods and people over long distances. This, in turn, stimulated trade and economic interconnectivity between regions.

The utilisation of petroleum-based energy systems, such as the internal combustion engine, played a pivotal role in powering various industries along with transportation. This newfound energy source enabled the development of automobiles, steamships, and other forms of transportation, resulting in increased mobility and trade on a global scale.

These developments not only fuelled economic growth but also laid the foundation for a more interconnected and globalised economic system. However, it is worth noting that some of the environmental challenges associated with industrialisation, such as pollution and greenhouse gas emissions, became more pronounced during this period due to the increased extraction and utilisation of coal and other minerals. This realisation about unsustainability of 'business as usual' and the need for enhanced energy efficiency has prompted the search for alternative technologies, particularly for the energy-intensive industries.

ENERGY-INTENSIVE INDUSTRIES

The industrial sector can be characterised as energy-intensive and non-energy-intensive depending upon the specific energy consumption, total energy usage and the impact on GHG emission. Energy-intensive industries are sectors of the economy that require a significant amount of energy to produce goods or provide services. These industries typically have high energy consumption relative to their economic

DOI: 10.1201/9781003415718-3

output. Energy-intensive industries play a crucial role in the global economy, but they are also major contributors to greenhouse gas emissions. Some of the key energy-intensive industries include mining, metallurgy, manufacturing, chemicals and petrochemicals, etc. Extracting minerals, ores, and fossil fuels from the earth's crust is highly energy-intensive, especially for deep mining operations and the transportation of heavy machinery. Metallurgical industries such as iron and steel production, aluminium smelting, and other metallurgical processes require enormous amounts of energy to extract and process raw materials. Other manufacturing industries such as cement, chemicals, glass and ceramics, and paper production are also among the most energy-intensive sectors. They rely on high-temperature processes and heavy machinery, which consume substantial amounts of mechanical energy. Chemical manufacturing involves various energy-intensive processes, including chemical reactions, distillation, and the production of petrochemicals. The textile industry, which includes the production of clothing, fabrics, and textiles, requires energy for processes such as dyeing, weaving, and finishing. Though specific energy consumption is low for textile manufacturing, overall impact is high due to volume effect.

Food, pulp and paper, basic chemicals, refining, iron and steel, nonferrous metals (primarily aluminium), and non-metallic minerals (primarily cement). Together, they account for about half of all industrial sector delivered energy use.[1]

As per IEA, the industry sector, including iron, steel, cement, chemicals and petrochemicals, accounts for over 30% of global total primary energy demand (IEA 2022). Iron and steel followed by refineries, cement, petrochemicals, fertilisers, lime and plaster, paper and pulp, and aluminium are considered the most energy-intensive industries accounting for over 80% of the GHG emission in Europe as per the EU-ETS in 2018 (Sander de BRUYN 2020[2]).

The top-1,000 industrial energy efficiency programme launched by China in 2006 included over 200 industrial units each in the iron and steel and chemical segments followed by over 100 units each in the power generation, petroleum and petrochemical and cement segments followed by number of units in other segments such as nonferrous metals, coal mining, pulp and paper, and textile segments. Many other industrial and non-industrial segments were further added in the top-10,000 programmes later.

India launched its 'perform, achieve and trade (PAT)' scheme in the year 2012 covering 478 designated consumers in eight industrial segments, aluminium, cement, chlor-alkali, fertilisers, iron and steel, and thermal power plants and textiles.[3]

Understanding the specific processes and energy systems of these energy-intensive industries is crucial for in-depth analysis of their energy consumption and implementation of energy-saving measures. The Author has reviewed several documents such as IEA reports on industrial energy efficiency, EU-ETS programme 2018, Top-10,000 programme of China and BEE PAT scheme of India. A shortlist of the industries has been prepared for detailed study with a view to set the context of energy efficiency opportunities in these sectors. The list includes:

Aluminium
Cement
Fertiliser

Iron and steel
Pulp and paper and
Textile.

ALUMINIUM

The aluminium-manufacturing process involves two main sub-processes: bauxite refining and smelting. Typical process and energy flow is shown in Figure 3.1.

Bauxite is the primary source of aluminium. Alumina is extracted from bauxite in the refinery. The refining process involves crushing and grinding of bauxite followed by digestion in hot caustic soda that dissolve the alumina from the ore materials. Impurities are settled and removed from the solution, leaving behind a clarified liquid. The clarified liquid is cooled and, in the process, alumina hydrate crystals are formed which are then separated out. The alumina hydrate is heated to high temperatures for production of alumina.

Alumina is then transformed into aluminium metal through the smelting process. The primary smelting method used is the Hall–Héroult process, which involves electrolysis. Alumina is dissolved in a molten cryolite electrolyte within a carbon-lined cell called a pot. Direct current is passed through the electrolyte, causing the alumina to decompose into aluminium metal at the cathode and oxygen gas at the anode. The molten aluminium metal accumulates at the bottom of the cell and is periodically siphoned off and further processed for converting this primary aluminium into various merchant products. This includes ingots as well as other products such as sheets, plates, foils, extrusions, and castings. Smelting accounts for over 70% of the energy consumption followed by other processes.

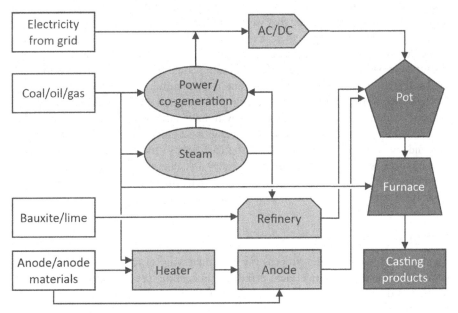

FIGURE 3.1 Process and Energy Flow – Aluminium.

The aluminium sector has been making large investment in technology and efficiency upgrade for reducing the specific energy consumption. Yet, there are still large number of underperformers with specific energy consumption in the refineries ranging from 10.2 to 24.5 GJ/t of alumina. An International Expert Committee has set the global benchmark at 7.80 GJ/t of calcined alumina. This benchmark serves as a reference point for evaluating the energy efficiency of refineries and encouraging improvements in energy consumption and carbon footprint reduction within the industry.[4]

The electrical energy consumption in the sector is primarily impacted by the voltage drops in the different sections of the electrolytic cell commonly referred to as 'pots' in the industry. The focus of energy efficiency is on reduction of the voltage drops adopting technologies for production of better anodes and cathodes, which will remain dimensionally stable over lifetime of operation. Technology of Inert anodes and wetted cathodes, currently under development offer several potential advantages over the traditional Hall–Héroult cells. In addition to reduced energy consumption, it also results in drastic reduction in greenhouse gas emissions. In the traditional process, consumable carbon anodes are used, which release carbon dioxide (CO_2) during the process of electrolysis. Inert anodes, on the other hand, do not degrade during the process and therefore do not emit greenhouse gases associated with the anode consumption.

Wetted cathodes, combined with the use of inert anodes, allow for decreased anode-cathode distances in the electrolysis cell. This reduced distance results in a lower cell voltage drop and consequently reduced energy consumption. The use of wetted cathodes can also contribute to lower operating temperatures in the electrolysis cell. Lower operating temperatures have several other co-benefits, including improved energy efficiency, reduced wear and tear on equipment, and potentially lower cooling requirements. It's worth noting that while inert anodes and wetted cathodes offer significant potential advantages, there are also technical challenges to overcome in their implementation. Developing cost-effective and durable inert anode materials, ensuring efficient wetting of cathodes, and addressing other operational considerations are areas of ongoing research and development in the aluminium industry.

Other initiatives for reduction of energy consumption and carbon footprint in the sector include increasing recycling thereby reducing the demand for virgin aluminium by up to 50%, high-efficiency cogeneration for meeting the power and steam demand, recovery of waste heat from the calciners in the refineries and furnaces. The industry is also heavily investing in renewable energy such solar and wind and technologies for integration of renewables. Green hydrogen is also being looked at for substitution of fossil fuel used in cogeneration facilities.

CEMENT

The cement manufacturing process involves quarrying of raw materials, crushing, pre-homogenisation and raw meal grinding, preheating, pre-calcining, clinker production in the rotary kiln, cooling and storing, blending, cement grinding, and storing in the cement silo.

Both wet and dry process technologies are there. However, wet process is no longer used commercially due to high energy and water consumption. In the wet process, raw materials (excluding gypsum) are crushed and mixed in an appropriate ratio. Water is added, and the mixture is further ground into a slurry with a water content of 35%–40% using a tube mill. The slurry is then stored in a tank, where it is homogenised with corrective materials, and finally sent to a rotary kiln for clinker burning. The wet process allows for easy mixing of the slurry, but it consumes a large amount of energy due to the evaporation of extra water during clinker burning.

In the dry process, crushed raw materials are dried in a cylindrical rotary drier, mixed at a predetermined ratio, further ground, and conveyed to different storage tanks. The prepared materials are mixed again according to a predetermined ratio and fed into a rotary kiln for clinkerisation. The dry process consumes less energy and has lower running costs compared to the wet process. The wet process has been phased out in all major cement producing countries due to high energy and water consumptions.

Both thermal and electrical energy are used in the cement manufacturing process as would be seen from Figure 3.2.

Thermal energy is used for calcination, which involves heating raw materials to high temperatures in the kiln to produce clinker. It is also utilised for preheating the raw materials and drying them before entering the kiln. This energy is usually obtained from the combustion of fossil fuels like coal, petroleum coke, or natural gas.

Electricity is used for various motor systems such as material transport, crushing, and milling of raw materials, as well as grinding of clinker and additives to produce cement as well as other utilities such as fans, compressors, pumps etc.

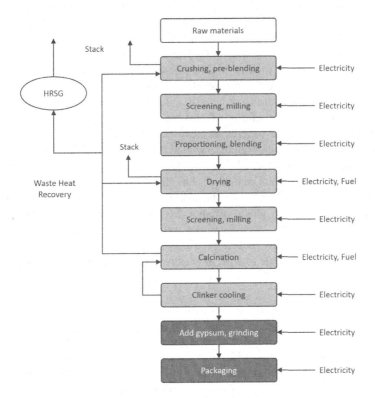

FIGURE 3.2 Process and Energy Flow – Cement.

Fuel constitutes about 90% of the overall energy consumption, while electricity accounts for the remaining share. The breakdown of electricity consumption shows that approximately 40% is consumed in clinker grinding, 30% in raw material processing, 25% in clinker production, and the remaining 5% in other auxiliaries and plant utilities.

Waste heat from the kilns provides a secondary energy source that are utilised for preheating as well heat recovery steam generation (HRSG) for power generation. Waste heat recovery is playing a major role in improving overall energy efficiency and reduction of GHG emission from the cement sector.

The cement industry has witnessed significant technological advancements in various equipment and systems, leading to transformations of the production process worldwide. Some of the key technology advancements includes precalcination, high-pressure grinding, high-efficiency particle separation, and digitalisation for process control.

The production process begins with pre-calcination in which a portion of the raw meal is preheated and calcined before entering the main kiln. This technology helps in reducing the fuel consumption in the kiln.

High-pressure grinding rolls (HPGR) have been introduced in cement plants as an alternative to conventional ball mills for raw material and cement grinding. HPGR technology offers advantages such as lower energy consumption, improved grinding efficiency, and better control of particle size distribution.

Innovations in clinker cooling systems have been introduced to improve the heat recovery efficiency thereby reducing the need for primary fuel of the process. Technologies like grate coolers and air quenching systems help in reducing energy consumption and enhance the quality of the final product.

Advanced separators, such as high-efficiency cyclones and dynamic air classifiers, have been developed to improve the particle separation efficiency in cement mills. These separators help to reduce the circulating load, increase mill capacity, and enhance the overall grinding performance.

In addition to the other benefits, cement industry is making substantial gains on energy productivity through digitalisation led process optimisation.

There is huge potential for reducing the carbon footprint in the cement industry by using alternative materials. The industry has already achieved 25%–30% blending of alternative materials such as fly ash, blast furnace slag, and municipal solid wastes.

Cement industry has a global alliance 'cement sustainability initiative',[5] which is a CEO led initiative by 24 major cement producers with operations in more than 100 countries for drastically reducing the carbon intensity of the sector by promoting energy efficiency, renewable energy, alternative materials, and carbon sequestration technologies.

FERTILISER

The invention of the Haber–Bosch process for production of synthetic fertilisers on a large-scale revolutionised the agricultural practices. Prior to this invention, organic fertilisers, such as plant and animal wastes, were the primary source of nutrients for crops. While organic fertilisers are still used, the ability to produce synthetic fertilisers has

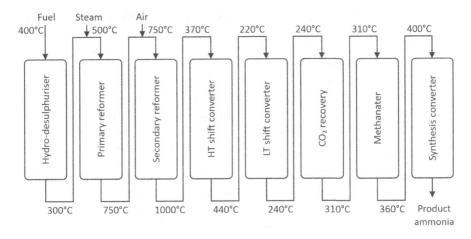

FIGURE 3.3   Process Flow – Ammonia Production.

greatly improved the efficiency and productivity of agriculture. Nitrogen is an essential nutrient for plant growth and plays a crucial role in protein synthesis, leaf and stem development, and overall plant vigour. Due to its importance, nitrogen fertilisers account for the largest share of the global fertiliser market, exceeding 60% of the market share.

The nitrogen in many straight and compound fertilisers is in the form of ammonium (NH_4^+ cation). The Haber–Bosch process is deployed for production of ammonia (NH_3) from atmospheric nitrogen (N_2) and hydrogen gas (H_2). H_2 is produced by steam reformation of fuel such as natural gas or electrolysis of water when cheaper and greener power is available. The process involves combining nitrogen and hydrogen under high pressure and temperature in the presence of a catalyst, typically iron or iron-based compounds.

The Haber–Bosch process requires high temperatures and pressures and utilises nitrogen fixation, the simplified process flow is as shown in Figure 3.3.

The process begins with hydrodesulphurisation, which involves reacting the sulphur compound present in the fuel with hydrogen at elevated temperature in presence of catalysts such as cobalt, molybdenum, or nickel. Hydrogen reacts with sulphur to produce hydrogen sulphide (H_2S), which is then separated out.

The primary reforming process, also known as steam methane reforming (SMR), is a widely used method for producing hydrogen gas (H_2) from natural gas. The purified natural gas is preheated to a high temperature, typically between 700°C and 900°C. This step helps in increasing the reaction rate and improve the efficiency of the reforming process. The preheated natural gas is then mixed with steam (H_2O) as per specified ratio. The steam acts as a reactant and a source of hydrogen in the reforming process. The mixture of natural gas and steam enters a reforming reactor, which contains a catalyst, typically based on nickel. The reactor operates at high temperatures, usually in the range of 800°C–1,000°C, and at elevated pressures. Inside the reforming reactor, a series of endothermic reactions occur.

In the secondary reforming step, the gas mixture from the water-gas shift reaction is further processed. It is heated and passed over a different catalyst, typically a mixture of nickel and alumina. The secondary reforming reaction also takes place at high temperatures (typically above 800°C) and converts any remaining methane and higher hydrocarbons into hydrogen and carbon monoxide. The reaction is endothermic, requiring considerable amount of heat input, which is provided by the input fuel.

High-temperature shift conversion, also known as the high-temperature water-gas shift reaction, is a chemical process to convert carbon monoxide (CO) and water (H_2O) into carbon dioxide (CO_2) and hydrogen (H_2) using a catalyst. The high-temperature shift conversion reaction is typically carried out at temperatures ranging from 300°C to 450°C. At these elevated temperatures, the reaction is exothermic, meaning it releases heat.

Low-temperature shift conversion, also known as low-temperature shift (LTS), is a chemical process used to convert CO into CO_2. The LTS reaction is a catalytic reaction that takes place using a catalyst, commonly iron oxide or a mixture of iron oxide and chromium oxide. The typical temperature range for this process is around 200°C–400°C.

There are few alternative methods for removal of carbon dioxide from reformed gas. These include chemical as well as physical method such as membrane filtration.

Methanation is the process of combining CO_2 with H_2 to produce methane (CH_4) gas, which is recycled for improving the yield of hydrogen from the steam methane reformation process.

N_2 and H_2 are reacted at high pressure and temperature in presence of iron-based catalyst, often combined with promoters such as potassium oxide (K_2O) or aluminium oxide (Al_2O_3).

As against about 9 Gcal per tonne of ammonia prevailing in 80s, currently gas-based plants are operating at near 7 Gcal per tonne of ammonia. The thermodynamic minimum energy requirement is 4.44 Gcal per tonne. However, the industry is finding it difficult to bridge the gap considering the current price of gas and investment required in newer technologies. Near-zero-emission technologies for ammonia production are emerging, including electrolysis, methane pyrolysis and fossil-based routes with carbon capture and storage (IEA 21).

Technologies for ammonia production are mostly licensed. As such, implementation of technology-oriented retrofit projects is always carried out by the licensors. The role of plant engineers' rests with assimilation of the technologies for carrying out operation and maintenance as per protocols. Ammonia plants have also energy-intensive off-site facilities that include cogeneration plants, steam network, large motor systems and hydraulic network, cooling towers, etc. The energy auditors and the plant engineers have a much bigger role in undertaking energy efficiency projects in these facilities following the audit and implementation methodologies prescribed in this book.

IRON AND STEEL

The iron and steel industry is a complex and diverse sector with various types of producers, production technologies, and geographic concentrations. As in case of cement, China plays a dominant role in the global steel industry. The country has a substantial number of state-owned integrated steel plants. These plants handle the entire steel production process from raw materials (iron ore and coal) to finished steel products. China's steel production capacity accounts for over 54% of the world's total steel production. Besides China, there are multinational steel companies operating large integrated steel plants in select countries. These companies have significant production capacities and are major players in the global steel market. Examples of such companies include ArcelorMittal, Nippon Steel, POSCO, and Tata Steel. In addition to the large integrated plants, there is a substantial number of smaller steel producers worldwide. These include producers using alternative methods such as sponge iron and electric arc furnaces (EAFs).

The top 10 producers in the industry account for over 25% of global production, which is relatively low compared to other sectors like aluminium. The iron and steel sector accounts for largest contribution to CO_2 emissions among heavy industries and second in energy consumption. The sector directly contributes to approximately 2.6 gigatonnes of carbon dioxide (Gt CO_2) emissions annually, accounting for 7% of the global total from the energy system (IEA Oct, 2020).[6]

The iron and steel industry operates in a global market with international trade playing a significant role. Different countries and regions have their own production capacities, demand patterns, and trade policies that impact the industry's structure and competitiveness.

Figure 3.4 shows the processes involved in making iron and steel at a glance.

The primary raw materials used in the iron and steel industry are iron ore, coal, and limestone. Iron ore is mined from deposits and then processed into iron through various techniques, such as sintering or pelletising. Coal is used as a fuel and a reducing agent in the production of iron, while limestone helps remove impurities during the iron-making process.

The manufacture of iron and steel involves several stages and processes. The primary process consists of two stages: the production of sponge iron and pig iron through reduction, followed by the production of crude steel. The secondary process involves the production of merchant products from the crude steel. The integrated manufacturing process typically utilises the blast furnace (BF) and basic oxygen

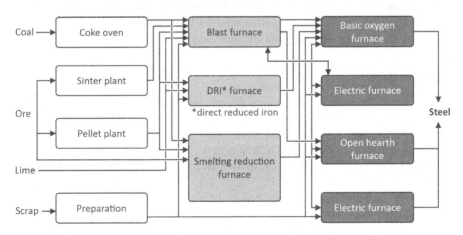

FIGURE 3.4   Steel Making at a Glance.

furnace (BOF), which are commonly used in larger integrated mills. Additionally, the direct reduction process (DRI) is employed for the production of sponge iron.

In the BF-BOF process, the first stage involves the reduction of iron ore to produce sponge iron. Originally, natural gas-based reformation technology was used in the DRI process. However, coal-based technology has been developed, and currently, most Chinese and Indian sponge iron units use coal as both fuel and reducing agent. Direct reduced iron, along with steel scrap, is then melted in an EAF to produce molten steel and subsequent products.

After the BF-BOF process or the DRI process, the molten steel is controlled to achieve the desired composition and temperature. It is then cast using a continuous casting machine to produce slabs, blooms, and billets. These castings are further processed by the rolling mill, where they are rolled to the required dimensions for making consumer products.

Iron and steel making require large amounts of energy to convert raw materials into finished steel products. The primary energy sources used in iron and steel production include coal, natural gas, petroleum coke, and electricity. Energy from secondary sources include BF and coke oven gas, waste heat from various furnaces. Iron making is by far the most energy-intensive process accounting for close to 60%–70% of the total energy consumed. Apart from the processes, the industry consumes significant amount of electricity in utilities such as auxiliaries for the cogeneration power and other plants.

The iron and steel industry has a global alliance 'world steel' for promoting breakthrough technologies for increasing energy efficiency and reduction of GHG emission.[7] These include:

Adoption of high end and proven technologies such as coke dry quenching (CDQ), BF top gas recovery turbines, waste heat recoveries from various furnaces

Implementing high-efficiency motor system retrofits, better refractory and insulating materials

Investing in research and development projects such as using green hydrogen as fuel, carbon capture, use and storage (CCUS) and integration with renewable technologies (solar, wind and biomass power systems)

PULP AND PAPER

The pulp and paper industry converts fibrous raw materials into pulp, paper, and paperboard products. Pulp mills manufacture only pulp, which is then sold and transported to paper and paperboard mills. A paper and paperboard mill may purchase pulp or manufacture its own pulp in house; in the latter case, such mills are referred to as integrated mills. Figure 3.5 shows the process and energy flow in an integrated paper mill.

Bulk of the requirement of raw materials is met from forest products, such as wood and bamboo. The logs undergo debarking to remove the bark since it can be a contaminant in the pulping process. The debarked logs are sent to a chipping machine which cuts them into smaller pieces known as chips as feed for pulping.

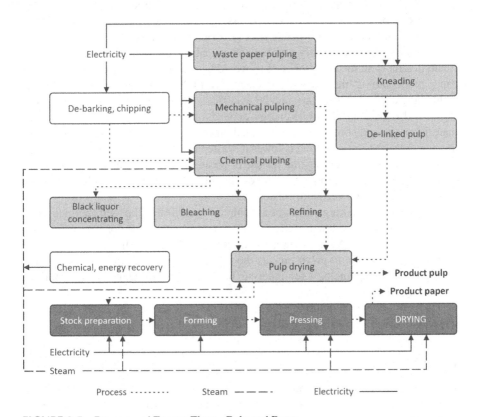

FIGURE 3.5   Process and Energy Flow – Pulp and Paper.

In addition to wood and bamboo, agro-waste-based mills utilise a variety of raw materials including stalks, straw, and sticks. These materials are also desized in cutters or shearing machines before use as feed for the pulping process.

Increasingly, paper mills are incorporating secondary fibre into their production processes. Waste paper is a major source of secondary fibre. This allows for recycling and reuse of paper products, reducing the demand for virgin fibres. This helps in reducing the energy consumption and GHG emission thereby promoting sustainability.

Pulping is the next process for extraction of fibre. Pulping frees the fibres in the feedstock from the lignin that binds these fibres together. The three main processes for producing wood pulp are mechanical pulping (includes thermo-mechanical process), chemical pulping (includes kraft and sulphite pulping also), semi-chemical pulping, and waste paper pulping. Mechanical pulping is the oldest form of pulping. The process employs mechanical energy to weaken and separate fibres from wood and waste paper feedstock via a grinding action. In the thermo-mechanical pulping (TMP) process, wood chips are first steamed to soften them before being ground in the same manner as the mechanical process. Chemo-thermo-mechanical pulping (CTMP) involves the application of chemicals to wood chips prior to refiner pulping.

The next stage is papermaking which is carried out in three-step process consisting of stock preparation, 'wet end' processing where sheet formation occurs, and finally 'dry end' processing where sheets are dried and finished.

The pulp and paper industry uses both electrical and thermal energy in the form of steam in almost all the sub-processes as would be seen from the Figure 3.5. Pulp and paper production is highly energy intensive with 60%–80% of the energy requirement being used as process heat and 20%–40% as electrical power (IEA 2020).[8] The share of thermal and electrical energy depends upon variable factors such as raw material usage and composition of finished products. Energy consumption for pulping and digesting, for example, is lower if wastepaper is used instead of wood chips or agricultural residue. In general, the use of wastepaper requires about 2.5 times less energy[9] than a similar production process based on other inputs mainly because of less intensive pulping needs for wastepaper.

Pulp and paper making processes account for over 70% of the total energy used in the manufacturing operation. Balance 30% is consumed mostly for various utilities and support systems (Authors energy audit database 2005–2020).[10]

The ratio of steam to power makes the industry ideally suitable for deployment of cogeneration technology for meeting both power and steam demand for the processes.

The pulp and paper industry offer maximum opportunities for energy efficiency intervention through retrofits for both the thermal and electrical energy.

TEXTILES

Textile industry is not energy-intensive industry if one considers the energy value add in the manufacturing process. However, the overall energy consumption in the sector and therefore, the GHG emission is very high because of the size of the sector. The textile and apparel industry account for about 4% of the total manufacturing energy consumption in China, and it is ranked as the sixth largest steam consumer amongst the 15 largest industrial energy consumers in the US. Textile industry has a diverse structure with various types of mills and factories involved in different processes. We have the composite mills and individual processing mills such as spinning, weaving, and wet processing. The energy profiles of these entities vary significantly.

Composite mills are involved in multiple processes from spinning to weaving and wet processing. They take in cotton bales as raw materials and produce finished dress materials. Composite mills typically require a significant amount of electrical energy for machinery operation, including spinning machines, and looms, and thermal energy as steam for wet processing such as dyeing and finishing machines.

Individual unit processing mills focus on specific processes like spinning, weaving, or wet processing. These mills perform only one of these processes and supply intermediate products to other mills or manufacturers. For example, a spinning mill produces yarn, which is then sent to a weaving mill to be made into fabric. The energy profile of these mills will depend on the specific process they undertake.

There are very large number of small-scale wet processing units operating in the informal sector across the developing world. These units specialise in processes like dyeing, printing, and finishing. Typically, these units have lower-scale operations and may use traditional or less energy-efficient equipment. However, it's worth noting that in recent years, there has been an increasing focus on promoting sustainable and energy-efficient practices in the wet processing sector.

Garment factories have gained prominence as a major extension of the textile sector. In terms of energy efficiency, garment factories can be likened to commercial buildings because a significant portion of their electricity consumption is attributed to lighting and space conditioning. The manufacturing processes in garment factories, such as cutting, sewing, and assembling, usually require less energy compared to the machinery-intensive processes in mills. However, the energy demand for lighting, ventilation, heating, and cooling systems in these factories can be substantial.

In this sector too, the core processing units such as spinning and weaving are carried out in highly automated machines under proprietary technologies. As such, motor systems outside of these proprietary islands offer opportunities for enhancing efficiency by adaptation of high-efficiency motor retrofit.

The wet processing units require considerable amount of steam for processing as well as drying. Significant amount of energy is saved by adopting high-efficiency boilers, cogeneration, and various other steam conservation measures. Micro-cogeneration technologies have been developed in China, India, and many other developing countries. Even the wet processing units in the informal sector are deploying these cogeneration units for reducing their overall energy consumption.

Wet processing is also involved in consuming large quantity of water. Reduction of water consumption reduces the energy consumption in the process and effluent treatment facilities.

Textile clusters are being promoted by Governments in many countries. Such clusters offer opportunities for deployment of co-operative steam or cogeneration systems for energy conservation as well as renewable technologies for reducing carbon footprints. Chauffage model is another way of improving the energy efficiency and reding carbon footprint. Small-scale utility companies invest in highly efficient cogeneration system and sell power and steam to individual manufacturing units in the clusters on long-term energy services contract.

Encouraged and assisted by the global companies engaged in the retail textile and apparel businesses and Governments, there is now increased focus on improving energy and water efficiency and reduction of GHG emission in the sector. Standardised technology packages have been developed under various technical assistance programmes that is helping in accelerated adaptation of the same by the industry. These packages include energy-efficient machinery, process optimisation, waste heat recovery, cogeneration, energy efficient insulation, and efficient lighting and HVAC. Energy efficiency consulting and EPC organisations can play a big role in implementation of retrofit projects under different modes.

ENERGY SYSTEM – ENERGY-INTENSIVE INDUSTRIES

Industrial operations require thermal energy for heating and cooling while electrical energy is required for various motor systems. Thermal energy accounts for much larger share of the energy consumption in the energy-intensive industries. Different grades of thermal energy are required for different types of process industries. High-grade thermal energy is required in the metal processing industries such as furnaces. Industries such as pulp and paper, textiles, food processing require low- to medium-grade heat for process heating. Waste heat from furnaces can be harnessed for production of steam and power utilising different types of waste heat recovery technologies. Industries hosting low-grade heat on the other hand offer opportunities for generating power by deploying high-efficiency cogeneration system. The efficiency of the thermal energy system depends upon how well we have managed the entire conversion process from fuel and raw material management to recovery of waste heat. In case of cement plant, for example, the fuel efficiency will be impacted by the moisture level in the fuel as well as raw meal and recovery and utilisation of waste heat from the different sources. Similar is the case with iron and steel industry wherein the energy requirement from external sources can be substantially reduced by utilising waste heat.

Textile and paper industries on the other hand require medium and low-grade energy as steam for various processes. It is possible to meet the demand for low pressure steam by installing low pressure boiler and sourcing power from the utilities. The investment cost for such a system will be low. However, by investing in high-efficiency cogeneration system, we can not only meet the power demand from the cogeneration but often generate surplus for exporting to the utility grid.

We need power for driving motor systems such as pumps, compressors, fans, conveyors in all types of industries. Each of these applications offer their own opportunities for optimisation of energy consumption.

We look at energy system in industries taking into consideration the individual elements such as conversion efficiencies and efficiencies derived by integration of the thermal and the electrical systems. In addition to the conventional energy audit, following few tools are used for systematic analysis of the energy system in industries and assessment of the minimum energy need for a particular process.

Process flow diagram
Energy balance
Heat and mass balance
Recovery of waste heat
Heat exchanger networking and pinch
Exergy balance

Process flow diagram provides us a picture of the energy and material flow for the process at a glance. Energy balance tool helps in quantifying the energy inputs and outputs of a system to determine energy losses and identify areas for improvement. Heat and mass balance is derived by analysis of heat transfer and mass flow within a system, allowing for a detailed understanding of energy and material flows. Recovery of waste heat study leads to identification and implementation of measures for recovery and utilisation of waste heat from various waste sreams, including high, medium, and low-grade heat sources.

Heat exchanger networking and pinch technology aims at optimisation of heat transfer network through identification of the pinch points (the minimum temperature difference) in a system.

Exergy is a measure of the quality of energy. This tool helps in analysing the quality of energy flows within a system and identification of waste streams for extracting more work.

Digital tools are available to facilitate the analysis of process efficiencies and the application of these tools. However, having a fundamental knowledge of the underlying principles is essential for effectively utilising these tools. This knowledge helps in interpreting the results, identifying potential areas for improvement, and implementing energy-saving measures in industries.

An illustrative example for each of these process analytical tools are shared in Figure 3.6.

Energy Balance

Energy balance is carried out for overall energy used in the process that is aggregate of fuel energy and electricity or individual components; fuel, electricity and secondary thermal streams such as steam. Figure 3.6 shows the electricity balance (KWH) for an integrated steel plant

Drawing up the energy balance makes it sure that every unit of the energy has been accounted for. From thereon, we can do various analysis while undertaking energy audit. The balance diagram also helps us in establishing the priority for energy efficiency studies considering the materiality and implementation constraints for the individual measures .

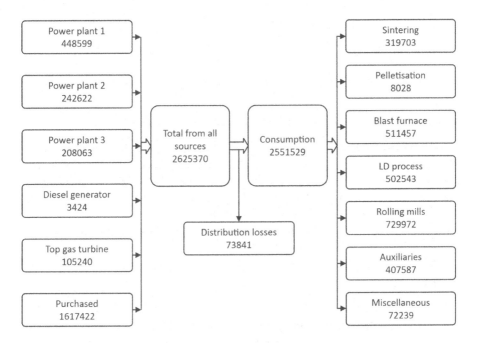

FIGURE 3.6   Electricity Balance – Integrated Steel Plant.

HEAT AND MASS BALANCE DIAGRAM

Heat and mass balance diagram (HMBD) shows how the energy gets converted as we proceed from the high temperature end of the thermal system to its end point. Pressure, temperature and mass flow rates are shown for every stage. HMBDs are extensively used for design of new projects as well as analysis of the existing steam power plants including cogeneration plants and waste heat recovery projects. The following is an example of HMBD for the steam and power cycle for a waste heat recovery project (Figure 3.7).

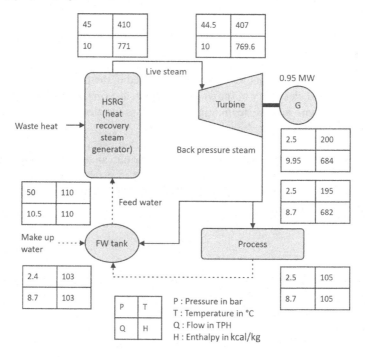

FIGURE 3.7 HMBD for a Waste Heat Recovery Cogeneration Project.

The figures in the boxes show the property of the fluids (pressure, temperature, flow, and enthalpy) during the recovery and conversion process. It is possible to prepare a single HMBD covering the furnace from where waste heat is generated and the process where the heat energy is used. However, it is better to do this island-wise to have a better idea on performance of individual islands. Spread sheet tools can be developed for real-time energy performance of the system by capturing the data and inputting the same to the individual tables.

RECOVERY OF WASTE HEAT

Waste heat recovery is getting increased focus as a source of energy in energy-intensive industries, particularly iron and steel and cement plants. A study conducted by Lawrence Berkeley National Labs estimated in 2005 that the US alone has 100 gigawatts of untapped electrical capacity in the form of waste heat that annually could produce 742 terawatt hours of power.[11]

Waste heat is a byproduct of various industrial processes where heat is generated through fuel combustion or chemical reactions. Instead of utilizing the same for useful purposes, this heat is often released into the environment. The value of waste heat lies in its potential to be reused for economic and practical benefits. The approach for recovering waste heat depends on the temperature of the waste heat and the economic feasibility of harnessing it.

In industrial processes, a significant portion of the heat generated is wasted due to inefficiencies in the heat-power cycle and the inability to utilise the residual heat in exhaust gases. Additional energy is often required to dissipate this residual heat into the environment. The amount and quality of waste heat produced vary across different industries depending on the specific processes involved.

Waste heat is classified into three categories based on temperature:

High grade: This refers to waste heat with temperatures exceeding 650°C. Conventional heat-power conversion technologies can be employed to harness this high-grade waste heat effectively.

Medium grade: Waste heat falls into the medium-grade category when its temperature ranges from 230°C to 650°C. Both heat-power conversion technologies and heating/cooling technologies can be utilised to make use of this waste heat efficiently.

Low grade: Waste heat with temperatures below 230°C is classified as low grade. Primarily used for heating purposes, this waste heat presents an opportunity for deploying innovative heat-power technologies such as the organic Rankine cycle.

The terminology 'waste heat' itself is a misnomer. A better expression will be 'thermal energy recovery potential'. Incredibly large number of technologies are there for thermal energy recovery from different systems.[12]

Table 3.1 provides an illustrative lists of energy recovery technologies.

TABLE 3.1
Thermal Energy Recovery Technologies

Waste Heat Quality	Source of Waste Heat	Technologies	Applications
High	Metal furnaces Kilns Incinerators	Heat recovery steam generators Gas turbine	Rankine cycle (steam to power) Steel plants
Medium	Gas turbine IC engines Furnaces Kilns	Vapour absorption machines for cooling Super chargers Gas to air heat exchangers Gas to water heat exchangers Gas to solid heat exchangers	Gas turbines (Cooling of charging air) IC engines (Compression of charging air) Air preheating for boilers and furnaces Process heating, drying Preheating of feeds, drying

(Continued)

TABLE 3.1 (*Continued*)
Thermal Energy Recovery Technologies

Waste Heat Quality	Source of Waste Heat	Technologies	Applications
Low	Boiler flue	Gas to air heat exchangers	Feed water preheating
	Furnace/engine cooling water	Gas to water heat exchangers	Process hot water
		Gas to air heat exchangers	Combustion air preheating
	Process contaminated condensate	Gas to water heat exchangers	Drying of fuel
		Organic Rankine cycle (ORC)	Micro power generation

Most of the technologies except for the Organic Rankine cycle (ORC) are now commercial. The energy-intensive industries provide both the source and sink for utilising waste heat that have helped in achieving large amount of energy savings through use of one or more these technologies.

HEAT EXCHANGER NETWORKING

Process energy consisting of heating and cooling accounts for a large share of the overall energy usage in most of the energy-intensive industries. We use this energy both directly and indirectly. In a cement kiln or in a metal furnace, heating energy is directly provided by the flue gas generated by combustion of fuel. In a boiler and many other processes, heat is transferred indirectly. We need heat exchangers for transfer of heat indirectly. A heat exchanger is an equipment utilised for transferring heat from one medium to another without letting the two mediums coming in direct contact. Apart meeting the cooling or heating need of the process, heat exchangers play a critical role in recovery of heat thereby maintaining high process energy efficiency. A series of heat exchangers are used for this. Heat exchangers networking is a diagrammatic representation of all the heat exchangers in a process separating out the heating and cooling streams. Cold streams enter the heat exchanger at a lower temperature, gets heated up in the exchanger and leaves at a higher temperature. Similarly, the hot stream gets cooled down by transferring the heat to the cold stream. The objective of heat exchanger networking is two-fold:

It ensures that the maximum amount of heat has been extracted from within the process for meeting the heating and cooling demand and reducing the external energy required for process heating and cooling

Minimum amount of energy will be required for disposal of waste heat for example vapour load from condensers

The principles behind HEN are explained with few simple diagrams as follows (Figure 3.8).

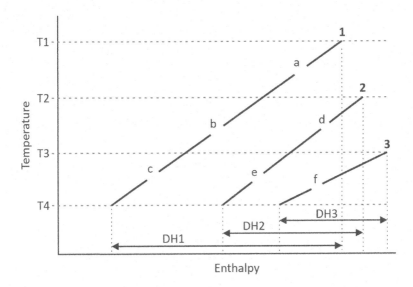

FIGURE 3.8   Hot Streams on Temperature Charts.

There are three hot streams 1, 2, and 3. Stream 1 is entering the heat exchanger at temperature T1, getting cooled down to temperature T4 loosing heat by DH1, consisting of a between T1and T2, b between T2 and T3, and c between T3 and T4. Similarly, the other two streams are also getting cooled down to T1 and loosing heat DH2 and DH3, respectively. All the three streams are giving up heat in stages between different temperature intervals T1and T2 (a), T2 and T3 (b + d), and T3 and T4 (c + e + f). We develop the hot composite curve (HCC) by plotting cumulative enthalpy drops between the temperature intervals. Similarly, we also develop the cold composite curve (CCC) by plotting cumulative enthalpy rise for each cold stream between the temperature intervals. It will be seen from the following figure that the curves are separated by varying degree of distance on both the scales. The point at which the curves are touching or having minimum distance on the temperature scale is the pinch point. In actual system, there can be multiple pinch points.

In the minimum energy design, the interchange of heat will take place between the HCC and the CCC on either side of the pinch. Violating this principle by having interchange of heat between the curves across the pinch point would invite double penalty. In that event, we will need more cooling and heating energy that the need for external energy support in Figure 3.9.

The principles behind HEN and Pinch are quite simple. But there are lot of complexities in applying these principles in practice. The whole idea behind including this section is to familiarise the readers of the book on the concept and arouse their curiosity. It would be beneficial to engage experts in the field for practical application.

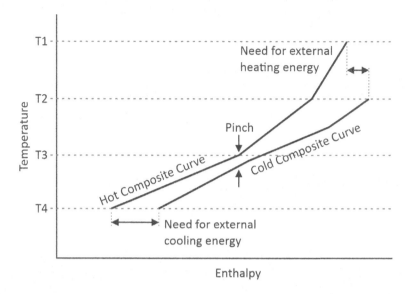

FIGURE 3.9   Hot and Cold Composite Curves.

EXERGY BALANCE

The second law of thermodynamics is a fundamental principle in physics that describes the behaviour of energy in systems. It provides insights into the direction of natural processes and the limitations on the conversion of energy from one form to another. While physicists often explain this law using concepts of reversibility and irreversibility, practicing engineers in industries may face challenges in applying these principles to analyse the performance of their energy systems.

In steam engineering, the properties of a thermal energy stream are typically characterised by parameters such as pressure, temperature, enthalpy, and entropy. Among these, engineers often find it difficult to comprehend the concept of entropy. Unlike other properties that can be directly measured, entropy cannot be measured directly but can be quantified through calculations based on other measurable properties. The concept of entropy becomes relevant in the context of available work. When the entropy level of an energy stream is lower than that of the surroundings or the destination of the stream, work can be extracted from it. The determination of available work can be accomplished using tools such as Mollier's chart (also known as the enthalpy–entropy diagram) or work equations.

Engineers need to establish a reference sink to assess the available work potential. This reference sink serves as a benchmark against which the available work or exergy of each energy stream can be evaluated. Exergy represents the portion of energy that can be converted into useful work, considering the system's reference state and the surrounding conditions.

P_o in Figure 3.10 represents the benchmark sink. There are two process streams, one at pressure and temperature P_1 and T_1 and the second at P_2 and T_2, respectively. The enthalpy of both the streams are same at H_1. The maximum work potential for the stream 1 is (H_1-H_0) and stream 2 (H_1-H_2) assuming isentropic expansion. Stream

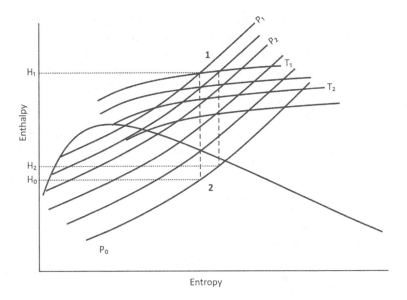

FIGURE 3.10   Determination of Exergy.

2 has less work potential, enthalpy remaining the same. This way we can determine the exergy content of every process stream to determine the loss of work potential.

The author has applied this concept real-time monitoring of the exergy level based on pressure and temperature of the various streams, outlet pressure and temperature from the boilers, inlet to the turbine and extractions and condensing steams. Huge financial gains have been achieved as a result

In summary, practicing engineers face challenges in understanding and applying the concept of entropy and not enough motivation to do so due to difficulty in assessing the gains. However, by utilising tools like Mollier's chart and establishing a reference sink, engineers can determine the available work or exergy of energy streams in their systems.

The author has included brief introduction on the various advance analytical tools, such as energy balance, heat and mass balance, recovery of waste heat, heat exchanger networking and exergy analysis that are used for energy efficiency assessment of energy-intensive industries. Each one of this subject is quite vast and this brief introduction should be considered by the readers as only to arouse their interest and attention. For further and deeper these concepts and their applications, the readers are referred to the bibliography.

CONCLUSION

The chapter on energy-intensive industries serves as a foundation for subsequent chapters on energy audits and the implementation of energy efficiency projects. has highlighted that these industries can make significant impact on GHG emission by reducing overall energy consumption. Therefore, it is crucial for industrial energy efficiency engineers to have a basic understanding of energy consumption in these industries, relevant efficiency technologies, and the tools available for identifying and implementing energy efficiency measures.

The selection of industries for inclusion in the list of energy-intensive industries is based on certain logical criteria, which are explained in the relevant section.

The chapter provides explanations of both process energy consumption (related to specific industrial processes) and utility energy consumption (associated with support systems such as motor systems, heating, cooling, and power generation). It also highlights potential areas for further study and analysis, which can be particularly useful for energy auditors.

The mentioned industries are just a few examples of sectors with distinctive energy-intensive processes. Gaining a basic understanding of the specific processes and energy systems employed by each industry is crucial for conducting energy audits and implementing effective energy efficiency measures. This knowledge enables a more comprehensive assessment and a tailored approach to identifying energy-saving opportunities within the unique context of each industry.

One important approach highlighted in the chapter is approach for minimum energy design and engineering. This approach is highly relevant for energy-intensive industries and aims to optimise energy efficiency throughout the design and engineering processes. The chapter briefly outlines various tools that can be used for minimum energy analysis. The application of these tools can significantly enhance the quality of energy audits conducted for energy-intensive industries.

In conclusion, the chapter on energy-intensive industries provides a necessary context for subsequent chapters on energy audits and energy efficiency project implementation. It emphasises the importance of understanding the energy consumption and specific processes within these industries and provides insights into potential areas for investigation. Additionally, it highlights the relevance of minimum energy design and engineering and introduces tools that can aid in conducting more comprehensive energy audits.

NOTES

1 Industrial sector energy consumption, (US EIA).
2 Sander de Bruyn et al., Energy intensive industries, challenges & opportunities in energy transition, 2020. https://www.researchgate.net/publication/342707922_Energy-
3 https://beeindia.gov.in/en/programmes/perform-achieve-and-trade-pat.
4 International benchmark energy consumption alumina bayer Kunwar, https://www.linkedin.com/pulse/international-benchmark-energy-consumption-alumina-bayer-kunwar/.
5 Cement sustainability initiative, https://www.wbcsd.org/Sector-Projects/Cement-Sustainability-Initiative.
6 IEA, Iron and steel technology road map, 2020, https://www.iea.org/reports/iron-and-steel-technology-roadmap.
7 World Steel, Breakthrough technologies, https://worldsteel.org/climate-action/breakthrough-technologies/.
8 The authors database.
9 Infographic-textile and apparel industries energy and water consumption and pollution profiles, https://www.linkedin.com/pulse/infographic-textile-apparel-industrys-energy-water-ali-hasanbeigi.
10 The author's database on energy audit.
11 Michael Kanellos, Will waste heat be bigger than sola, January 24, 2009, https://www.greentechmedia.com/articles/read/will-waste-heat-be-bigger-than-solar.
12 John L. Boyen, *Thermal Energy Recovery* (Hoboken, NJ: John Wiley & Sons, 1980).

4 Discovery of Opportunities
Energy Audit and Diagnostics

INTRODUCTION

We are all familiar with the term 'audit' used by the financial accounting professionals. The auditors scrutinise the books of accounts and certifies the same against the compliance requirement. The auditor has to verify that the accounts of the organisation are accurate and represent a true and fair picture of the financial status of the company. Also, the auditor must ensure that all material information has been recorded in the accounts. In case of any discrepancy, the same is highlighted by the auditors with qualifying remarks but without any recommendations on corrective action. That is where energy audit differs from financial audit. The energy auditors are required to identify the sources of inefficiency and also provide recommendations on measures for reducing inefficiencies.

Let's consider another parallel from the medical profession. Doctors makes some initial diagnosis based on feedback from the patient and prescribes some medicines. Often that is just the beginning of their interventions. Depending upon the progress, they conduct further diagnostic studies to identify the root cause of problems before prescribing more effective intervention measures. Taking a cue from the two examples cited above, this chapter on energy audit has been named 'Discovery of Opportunities-Energy Audit & Diagnostics' underscoring the importance of diagnostic studies for more impact making energy audit. We will continue with the term 'energy audit' as followed globally. However, we will demonstrate the usefulness of combining diagnostic tool such as root cause analysis for making the discovery process more impactful.

According to Wikipedia, an energy audit is defined as 'an inspection survey and analysis of energy flows for energy conservation in a building'. We can replace the term 'building' with 'facility' to include industrial settings. Several similar definitions of energy audits exist. In its guidebook, Canadian Industry Partnership for Energy Conservation (CIPEC)[1] defines energy auditing as 'a systematic, documented verification process of objectively obtaining and evaluating energy audit evidence, in conformance with energy audit criteria, followed by communication of results to the client'. In the Indian Energy Conservation Act of 2001, an energy audit is defined as 'the verification, monitoring, and analysis of energy use, and the submission of a technical report containing recommendations for improving energy efficiency, along with cost-benefit analysis and an action plan to reduce energy consumption.' This definition seems to be closer to the concept of 'energy diagnostic study' deliberated above. By appropriately

DOI: 10.1201/9781003415718-4

adding the terms 'measurement' and 'reviews', we can make it more comprehensive. Consequently, the definition can be rephrased as 'Review, measurement, verification, monitoring, and analysis of energy use, along with the submission of a technical report containing recommendations for improving energy efficiency, including cost-benefit analysis and an action plan to reduce energy consumption'.

The term 'review' has been added to include evaluation of the current energy management system in a facility as part of the audit process. This aids in better defining the scope of the audit. We will delve further into this process as we unfold this chapter. Likewise, the term 'measurement' has been included since auditors are required to collect data through field measurements in addition to reported data, allowing for data validation and filling data gaps.

In essence, a high-quality energy audit is a comprehensive process of discovery that covers every aspect of energy management within a facility. Figure 4.1 illustrates how a well-structured discovery process could look like.

The proposed energy audit process shown in Figure 4.1 is like delayering of onion peels, the deeper one goes more layers get unfolded. This process was developed as a concept by the author while preparing a long-term energy efficiency programme in a chemical and fertiliser manufacturing facility. The investigation process advances from the outer layer to the next inner one revealing the opportunities for energy efficiency measures residing in that layer. This process enables an organisation to make an impact

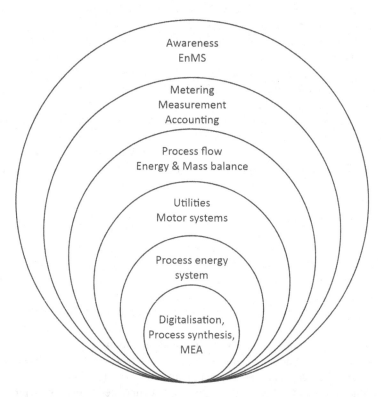

FIGURE 4.1 Unfolding the hidden opportunities.

making intervention for saving energy irrespective of the current technology status. The process can begin with analysis of simpler part of the energy system like lighting, HVAC, industrial motor systems for which the audit processes are fairly standardised. One moves from there to cover other areas of utilities such as boilers, furnaces, power generation, and cogeneration. We require higher level of skill competency for this.

Complexity increases as we go down to the next level of audit of the process energy systems. Every industrial process has its own unique characteristics, often based on patented technologies. The process heating and cooling system in an ammonia plant is entirely different from say, a paper manufacturing plant. The auditor therefore, must have a good amount of domain knowledge in addition to knowledge of process energy system in general for conducting process energy audit. This requirement increases manifold as we move deeper into the process. The exploration process involves application of state of art technologies such as process integration, heat exchanger networking, synthesis of control systems, digitalisation, artificial intelligence for assessing the minimum energy need. Obviously, the cost for such audit will be high due to requirement of higher level of knowledge and skill competency and time required for carrying out such analysis. But more often than not, benefits will far outstrip the cost.

The benefit from an energy audit for a facility also depends upon the preparedness of the facility engineers to participate in the audit process. Much larger benefit (higher yield-to-efforts ratio) is usually derived in a facility that is well aware of energy efficiency technologies and having in place a structured energy management system such as EnMS 50001.

For an external auditor, the first challenge is to assess the level of awareness about energy efficiency amongst the managerial and engineering professionals? This is best done by following a structured interview with the facility engineers without making it look like an interview. This interview process helps in freezing the scope of the energy audit. This also helps in assessing the skill competency requirements for the audit and the extent of support from the facility operation and maintenance engineers.

This chapter on energy audit has been prepared as a practical guide to enable both internal and external auditors to carry out different kinds of audits covering simple technologies such as the motor systems to the very complex one such as minimum energy approach. The chapter provides information on different types of energy audit and their applications followed by more elaborate descriptions of frequently used tools such as 'walk-through audit' and 'detailed energy audit'. Summary reviews have also been provided on the 'minimum energy approach'.

Stakeholders' engagement is very important for successful conducting of audit and implementation of identified energy savings measures. Section on the stakeholders' engagement process has been substantiated by an interesting case study from the MSME industrial segment.

ENERGY AUDIT TYPES AND APPLICATIONS

Different types of energy audit have been developed over time by the audit fraternity across the globe. However, we do not yet have defined global standards for different types of audits, particularly for industrial facilities. The classification provided by ASHRAE for commercial buildings can be broadly adopted and adapted for industrial energy audits.

ASHRAE's classification includes three levels of energy audits: Level 1, Level 2, and Level 3. These levels can be applied to industrial energy audits as well. Here's a brief description of each level:

Level 1 Audit (Walk-through Survey): This initial survey involves a preliminary assessment of the facility's energy performance. It typically includes an analysis of utility bills, interviews with facility engineers, and identification of obvious energy wastages. Level 1 audit is primarily conducted to gather information and prepare a scoping document for a more detailed energy audit.

Level 2 Audit (Detailed Energy Audit): Level 2 audit involves more in-depth data collection and analysis. It includes the collection of usage data, detailed analysis of energy consumption patterns, and identification of energy-saving projects that can be implemented based on the available data. This level of audit aims to identify feasible energy efficiency measures without requiring extensive further data collection.

Level 3 Audit (Comprehensive Audit): Level 3 audit goes beyond the Level 2 audit by conducting additional data collection using sub-metres, portable metres, or other measurement devices. This level of audit involves more rigorous technical and financial analysis, particularly for capital-intensive energy projects. The purpose is to validate the findings from the Level 2 audit, perform detailed calculations, and conduct a comprehensive assessment of the proposed energy-saving measures.

ASHRAE has also included another terminology called targeted audit, which is followed for benchmarking and carrying on the audit to bridge the gap against benchmark.

While these levels provide a framework for conducting energy audits, it's important to note that the specific methodologies and depth of analysis may vary depending on the industry, facility type, and the goals of the audit. Customised approaches are usually required while scoping for addressing unique energy characteristic of a particular facility and demand from the facility owner.

The Canadian Industry Program for Energy Conservation (CIPEC) is a partnership between Canadian Government and industry for promoting innovative energy management in the participating industry for increasing profitability, sustainability and competitiveness (NATSOURCE, Government of Canada), Like many other such programmes across the globe, it has published guidelines (CIPEC 2002) and audit tools (CIPEC 2009) on two different levels of audit, preliminary audit (walk-through audit) and detailed audit (Diagnostic audit). The detailed audit scope as per this guideline combines the level 2 and 3 audits methods as per ASHRAE standards.

Practically, energy audit consists of a two-stage process, walk-through audit for preliminary assessment followed by detailed audit for identification of specific energy-saving measures. However, within the scope of detailed audit, we should remain open to provide some common as well as differentiated audit services to meet specific requirements of a particular facility. A comprehensive list of different audit types has been prepared based on over three decades of multi-country and multi-industrial facility auditing experience of the author as shown in Table 4.1.

TABLE 4.1
Different Types of Energy Audits

Audit Types	Key Characteristics	Applicability
Walk-through audit	Review based on client data and short field visit	Scoping for detailed energy audit
Compliance audit	Validation of information provided by facility and preparation of compliance report	Complying with the demand from the regulatory authorities as per prescribed formats
Detailed energy audit	Detailed investigation of the energy system, identification of energy savings measures, configuration of projects with cost-benefit analysis	Every where
Diagnostics study	Deep dive for root cause analysis	Specific for a particular system or an equipment as add-on to the detailed audit
Investment grade audit	Various options for financing, detailed analysis and preparation of bankable project reports	For highly capital-intensive projects such as a cogeneration or a waste heat recovery system
Audit for energy services performance contract (ESPC)	More detailed analysis of baseline energy consumption, baseline adjustment methodology and M&V plan	ESPC by ESCOs
Minimum energy approach	Top-down investigation based on pre-set target, digitalisation	Energy-intensive process industries aiming for higher levels of energy savings

Knowledge of each of the audit-type outlined in the table helps in preparing a better scoping document for the detailed energy audit and estimation of the audit cost.

WALK-THROUGH ENERGY AUDIT

Walk-through energy audit is carried out to make a preliminary assessment of the current energy performances of a facility and preparing a scoping document for detailed energy audit. The assessment is carried out by desk analysis based on the information provided by the facility followed by a short visit to the facility. The information provided by the facility is validated by checking the source of data and the review of the data acquisition system. Sometime low-cost portable instruments are used to capture additional data while going around the plant. The examples are temperature survey for insulated furnaces, boilers, vessels, pipelines, thermography for electrical panels, etc. These are done on sampling basis to sensitise the facility engineers on the current situation about energy losses and need for detailed study for quantification of the same.

The process of energy audit is fairly standardised in case of buildings and to some extent for industrial utility systems. But the same cannot be said about the process energy system due to the nature of diversity of various industrial processes and the scale of energy consumption. Bulk of the energy consumption in a cement plant is on account of fuel used in the kiln for production of clinker. A walk-through audit of the kiln can be performed with very little effort. However, information gathered during the walk-through audit is unlikely to be of much use for determination of scope for detailed audit. The situation changes if we take the case of paper industry. Pulp and paper production like cement is also energy intensive with 60%–80% of the energy requirement for process heat and 20%–40% for motor systems and other electrical appliances.[2] Within a particular sector, the share of thermal and electrical energy can vary considerable depending upon the use of raw materials and type of finished products. The energy usage, both composition and quantum-wise, in a waste/recycled paper-based mill producing kraft paper will be quite different from a wood-based mill producing news-print. The level of efforts required for detailed energy audit will be much lower in case of the waste paper-based mill. We generally adopt a three-step process even for walk-through audit, planning for the audit, field assessment followed by analysis and preparation of proposal for detailed audit.

PLANNING

The planning process involves preparation of a questionnaire, obtaining preliminary information from the facility, review of the quality of information and a benchmark analysis based on the auditor's database. The resource requirement and the organisation for carrying out different tasks are determined. A technical proposal is prepared on the scope of the audit. A budgetary cost proposal is also prepared accordingly. Deliverables at the different stages of the audit are mutually agreed. A non-binding memorandum of understanding (MOU) is signed before undertaking the next step of field visit.

FIELD VISIT

The main purpose of the field visit is to validate the information provided, freezing the scope of the audit, work schedule and finalisation of the cost budget.

PROPOSAL FOR DETAILED AUDIT

The process of walk-through audit concludes with submission of proposal for detailed energy audit covering the audit process, target for energy savings, cost budget, and the roles and responsibilities of the parties. In most cases, this leads to signing of the contract for detailed audit subject to the quality of the report and the value as perceived by the facility.

DETAILED ENERGY AUDIT

The process of detailed energy audit is well known amongst energy professionals. Number of very well-written documents and books are available on the subject in the public domain. However, as mentioned above, we still do not have a standard protocol for this. Further, the value perceptions about energy audit is changing with rapid

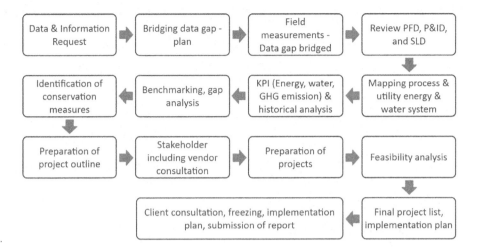

FIGURE 4.2 Detailed energy audit at a glance. KPI:Key performance indicator, PFD: Process flow diagram, P&ID: Piping and instrumentation diagram, SLD: Electrical single line diagram.

development of the energy and environmental scenarios and energy savings technologies. A comprehensive approach can be adopted for the detailed energy audit to respond to these changing scenarios as outlined in Figure 4.2.

Contents of the individual processes are well known to established energy auditors. Considering the focus of this book on implementation, the author would like to invite the attention of the readers to few essential pre-requisites for improving the quality of audit as follows.

AUDIT ORGANOGRAM

The energy audit process in an energy-intensive industry requires a dedicated and knowledgeable audit team with defined roles for each member. Creating a unit-specific audit team with expertise in unit processes, process energy systems, measurement, analysis, and client relationship management is crucial.

The energy audit process in an energy-intensive industry is quite resource intensive. The audit team must have good knowledge of the unit processes as well as the process energy system in addition to the measurement, analytical, and client relationship management skill required for conducting the audit. It is therefore, important to create a unit-specific audit team with defined roles for the individual members. A typical organogram may look like as illustrated in Figure 4.3.

FIGURE 4.3 Audit organogram.

Digitalisation has considerably reduced the time required for conducting energy audit. It is possible to carry out some of the activities in parallel mode and often without human interface. Much of the data can be directly transmitted to the auditors back office subject to the access provided by the client to its data acquisition and archival system and real-time data transfer capability of the portable instruments. Bulk of the analytical work can be carried out using home-grown or commercially available software's. However, this process requires careful handling to avoid duplication of efforts and quality problem. It is useful to have quality check system for all the audit processes. This generally helps in higher level of credibility with clients. The cost of audit also reduces due to avoidance of rework.

SAMPLING FOR FIELD MEASUREMENTS

Field measurements are conducted for evaluating energy performance of equipment and systems. The cost of audit is directly proportional to the number of such tests. We are therefore, confronted with the questions of how large should be the size of the samples, what should be the points of measurements. Generally, we decide the number of samples intuitively and in consultation with the client's engineers. We can use statistical tools for this.

Both probability and non-probability methodologies are practiced for sampling. However, for our kinds of projects, we would use probability methodology as we have access to data on population and their characteristics. For this, there are several approaches such as random sampling, stratified sampling, systematic sampling, and cluster random sampling. Stratified sampling suits us best. In fact, we are already doing it without specifically mentioning this in a structured manner in our reports. A sub-section can be provided in the report explaining the rationale of stratification and the strata that have been identified for sampling, testing, analysing, and interpretation of results.

Most of the appliances can be covered under two or three strata for the purpose of energy audit:

By type (Pumps, fans, compressors, heat exchangers)
By capacity (kW, M³/h)
By vintage (<5 years; 5–10 years, >10 years)

However, logic for the stratification should be established in consultation with the client.

Similarly, there are standard methodologies for deriving the number (%) of samples that should be covered for statistical validity based on statistical terminologies such as total population, confidence interval (CI), confidence level (CL), and response percentage. Only in case of a very large population, we will use such statistical model. For all practical purposes, 10%–15% samples per strata suffices in meeting the objective of the audit.

Developing Key Performance Indicators

According to ISO 50001, 'energy performance indicator (EnPI, though we are using the expression KPI in this book) is a quantitative value or measure of energy performance, as defined by the organization'.[3] The importance of KPI in creating awareness on energy efficiency was highlighted with an illustrative case example in Chapter 2 on Fundamentals of Energy Efficiency and Conservations. Development of KPIs follow the sequence of operation in the facility driven up from the individual equipment to the unit operations to the final output. Let us take the case of a limestone preparatory section of a cement plant as illustrated in Figure 4.4.

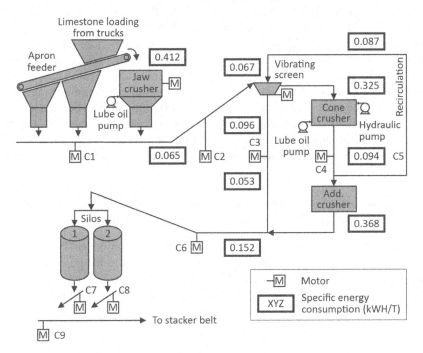

FIGURE 4.4   Drilled down SEC.

Specific energy consumption figures (figures in the boxes kWh/T of limestone) have been worked out for all the equipment in the section including the larger ones such as the crushers and smallest of the conveyors. Taking these figures as the baseline energy consumption, it is now possible to set different sets of KPIs:

Benchmarking SECs
Measurable improvement targets for individual equipment
Targets for individuals involved in operation and maintenance

This creates the environment for establishing accountability at various levels to manage the energy system in an efficient manner. KPIs of individual sections and individuals responsible for energy management creates the motivational force for driving both the efficiency and conservation measures, more so the later. While efficiency measures require investments, conservation through improved operation and maintenance often yield much higher benefits without much investments. This is what can be gained from a high-quality energy audit as would be seen from the case illustration on root cause analysis in the next section.

Deep Diving – Root Cause Analysis

A thorough analysis of energy-consuming systems and equipment is essential to identify the root causes of energy inefficiencies. Deep diving into the operational parameters, control strategies, maintenance practices, and other relevant factors can reveal opportunities for huge energy savings. This analysis involves studying equipment performance data, conducting on-site inspections, and engaging with facility personnel. KPI-based monitoring generates information on actual performance against target. Corrective measures are taken to bridge the identified performance gaps. The diagnostic study for identification of root cause of poor performance is relatively simple for motor systems but can be quite complex for equipment such as boilers, furnaces and process systems such as heat exchangers, distillation columns, evaporators, digestors, dryers, waste heat recovery systems, etc. This is so because of multiple variables that impact the performance of such equipment and systems. Let me illustrate this with a case study on a biomass-fired industrial boiler.

Case Study 4.1: Boiler Performance Diagnostics

This case pertains to a 50 TPH biomass-fired spreader stoker boiler. The boiler was designed to operate at 80% efficiency depending upon the moisture content of the fuel. The boiler was able to achieve efficiency ranging from 70% to 75% leaving a gap of about 5%. Various losses were computed by undertaking an indirect efficiency test following the ASME PTC-4 standard (Table 4.2).

TABLE 4.2
Results of Boiler Efficiency Test

SN	Different Boiler Losses (%)	Elements	Target	Actual	Gap
1	Dry flue gas	L1	5	6.28	1.28
2	Moisture from hydrogen	L2	5	7.87	2.87
3	Moisture in fuel	L3	4	3.28	-0.72
4	Moisture in combustion air	L4	0.2	0.20	0.00
5	Incomplete combustion (CO formation-Assumption)	L5	0.1	0.11	0.01
7	Unburnt carbon in bottom ash	L6	2	3.80	1.80
8	Unburnt carbon in fly ash	L7	1.5	2.24	0.74
9	Radiation and convection (Assumption)	L8	1	1.50	0.50
10	Total losses %	L	18.8	25.28	6.48
11	Boiler Efficiency	%	81.2	74.72	6.48

Amongst the controllable losses, unburnt carbon losses aggregate of loss through bottom and fly ash is higher by over 2.5% followed by 1.28% on dry flue gas account. Loss on account of moisture from hydrogen combustion, which is also higher by 2.87% is not controllable as it is a function of hydrogen content in the fuel.

Loss due to unburnt carbon is controlled by maintaining air–fuel ratio and ensuring proper mix of fuel with air. Increasing the air–fuel ratio reduces unburnt carbon but would lead to increase in dry flue gas loss, which is already higher. Higher loss on both carbon and dry flue gas account is indicative of improper mixing of air and fuel and lower residence time for combustion gases in the furnace due to higher gas velocity. We therefore, decided to carry out a more detailed investigation of the distribution of air flow and fuel spread in the furnace.

For flow mapping and analysis, the grate was divided into nine equal sections to form 81 grids for flow measurement. This was done to better understand the impact of:

Damper operation and position on air flow distribution and regulation
Repair condition of travelling grate fire bars on the flow pattern

FD fan was started and dampers adjusted to replicate the position corresponding to full load of the boiler. Air flow was measured using an anemometer over the 81 grid points. Figure 4.5 shows the patterns of the distribution.

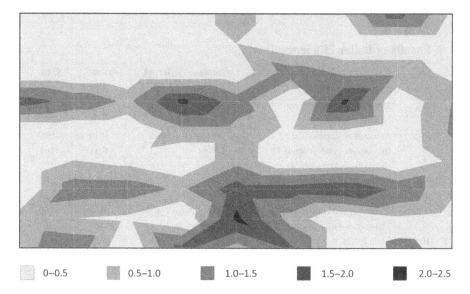

| 0–0.5 | 0.5–1.0 | 1.0–1.5 | 1.5–2.0 | 2.0–2.5 |

FIGURE 4.5   Air velocity (m/s) distribution over grate.

This shows that over 50% of the grate area is starved of air, velocity remaining less than 1 m/s against the calculated requirement of about 2 m/s. There are several spikes, velocity exceeding over 2.5 m/s. Overall average is about 2 m/s as per requirement. The grate bars and the assembly were examined. In addition to the very obvious problem of non-uniform air gaps as seen from Figure 4.6, many of the air holes were found blocked due to depositions.

FIGURE 4.6   Non-uniform gaps over the grate.

In addition to the adverse impact on combustion due to non-uniform air distribution, other negative impacts include localised overheating causing formation of clinker and burning of grate bars.

Corrective actions included modification of the air plenum chambers and flow control dampers, repair of the travelling grate to the maximum extent possible considering the high cost of replacements. Spot checks were carried out to assess the impact of rectification measures with the air dampers in fully open position (Table 4.3).

TABLE 4.3

Air Distribution Post Rectification

Left End	Centre-1	Centre-2	Centre-3	Centre-4	Right End
$V=2.9$ m/s	$V=3.1$ m/s	$V=2.9$ m/s	$V=2.9$ m/s	$V=2.8$ m/s	$V=2.7$ m/s
$F=855$ m³/min	$F=800$ m³/min	$F=605$ m³/min	$F=843$ m³/min	$F=744$ m³/min	$F=710$ m³/min
$V=3.3$ m/s	$V=3.0$ m/s	$V=3.2$ m/s	$V=2.9$ m/s	$V=2.7$ m/s	$V=2.9$ m/s
$F=740$ m³/min	$F=820$ m³/min	$F=740$ m³/min	$F=655$ m³/min	$F=650$ m³/min	$F=599$ m³/min
$V=3.1$ m/s	$V=3.3$ m/s	$V=3.3$ m/s	$V=2.9$ m/s	$V=2.7$ m/s	$V=2.9$ m/s
$F=855$ m³/min	$F=710$ m³/min	$F=610$ m³/min	$F=550$ m³/min	$F=655$ m³/min	$F=566$ m³/min

V, velocity; F, flow; m/s, metre per second; m³/min, cubic metre/minute.

Though much improved, there were still some pockets of spikes due to repair conditions of some of the links.

Similar studies were carried out to check the uniformity of the biomass spread. Biomass feeding arrangements consisting of fuel chutes from silo, mechanical feeders followed by pneumatic spreaders for throwing the fuel into the combustion space in the furnace. Mechanical and adjustable deflector plates are there for ensuring uniform distribution of the fuel across the grate width. The kinetics of the fuel flow is shown in Figure 4.7.

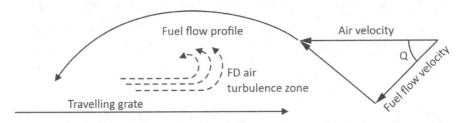

FIGURE 4.7   Fuel flow kinetics.

In a front discharge spreader stoker, the fuel distribution needs to be maintained to develop a profile starting with maximum flow near the rear wall gradually tapering down towards the discharge end. This ensures:

- Complete combustion due to increased retention time in the furnace
- Uniform heat flux
- Adequate cooling of the ash before discharge ensuring minimum loss due to sensible heat

Quantity of fuel feed to the individual aperture of the furnace is adjusted as per steam demand by varying the speed of the feeders. The air pressure at the spreader nozzles and the tilt angle of the deflector plates are adjusted to shift the peak towards the rear gradually tapering off towards front.

Figure 4.8 shows the fuel spread during the trial operation before corrective action.

FIGURE 4.8   Non-uniform fuel spread.

There were many peaks and valleys indicating improper functioning of the spreaders as well as the deflector plates. The corrective actions included repair of the spreader air-nozzles and replacement of the worn-out deflector plates. Figure 4.9 shows the position post corrective actions.

FIGURE 4.9   Fuel spread post corrective action.

By implementing improvements in the combustion system of the boiler by reducing unburnt carbon losses and optimising excess air requirements, the efficiency of the boiler was improved to approximately 77%. However, further improvements were expected after repairing the grate and fine-tuning the fuel feeding and spreading system.

It is acknowledged that conducting an exhaustive diagnostic study requires more effort and expertise compared to a regular energy audit. However, the potential rewards in terms of lower fuel consumption and improved uptime performance of the boiler are also significantly higher. The cost of the study is expected to be minimal compared to the annual savings achieved.

The key questions are, what areas need to be investigated, how to conduct the investigation, and what expertise is required. The investigation process remains the same regardless of the technology involved. It is essential to have domain specialists who possess expertise in the specific technology being investigated. Organisations that have implemented ISO 50001 energy management systems will find it easier to carry out such investigations. Technique such as total productive maintenance (TPM)[4] is another very useful tool for such analysis and managing the process of continuous improvement.

Financing

The energy audit recommendations will be incomplete without financial analysis of the projects and evaluation of options for financing. Assessing the feasibility of proposed measures, estimating costs and savings, exploring available financing options, and calculating payback periods are crucial steps in the audit process. This helps facilitate decision-making and supports the implementation of energy efficiency projects. More details on financing of retrofit projects have been provided in Chapter 5 on implementation.

COMPLIANCE AUDIT ENERGY, EMISSION

Energy-intensive industries are generally regulated in some form or other in many countries. Such regulated entities are required to periodically carry out different types of energy audit, implement energy efficiency measures and report the same to the regulators. The auditors are required to prepare and submit the audit reports as per standard formats in case of compliance audit whereas there is no such requirement in case of detailed energy audit. Japan had introduced energy management regulations for designated industrial consumers early in 1999. The designated entities are obligated to institute energy management system, appoint energy managers who are required to submit reports periodically to the regulators.

China had introduced the most intensive reporting requirement under different types of energy audit (Detailed mandatory audit, Quality assurance audit, Validation audit), under the Top-1,000 and later on Top-10,000 programme on energy efficiency.[5]

Bureau of Energy Efficiency in India operates a scheme called 'Perform, Achieve & Trade (PAT)' scheme for the industrial sector. PAT is a regulatory instrument to reduce the 'Specific Energy Consumption (SEC)' of products manufactured by designated industrial units (designated consumers, DC) in energy-intensive industries.[6] The DCs are mandated to carry out energy audit by external and empaneled energy auditors for validating the reported SECs by the DCs. The auditors are required to submit their report on the validation process as per approved formats.

In case of compliance audit, the auditors are normally not required to carry out any investigative work except for validating the data and information submitted by the facility. However, the auditors are expected to check the integrity of the energy metering, measurement and accounting system based on which the energy performance report has been prepared.

Increasingly, more and more companies are required to take action on reduction of GHG emission as per the country specific net zero plan and submit report on compliance. There are expert agencies for carrying out carbon audits. Even then, the energy auditors are expected to have the basic minimum knowledge of carbon accounting system.

Carbon accounting framework is used measuring and tracking amount of GHG emission at different levels, process, products, entity, and sector country. An individual entity establishes the emission baseline and sets targets for reduction of the same through adaptation of energy efficiency and renewable energy. The accounting method helps in managing the process in a transparent manner following global standards. A variety of standards and guidelines have been developed and codified under Greenhouse Gas Protocol and ISO 14064 for accounting at both product and project levels.

GHG emissions are generally accounted at three levels, scopes 1, 2, and 3. The Scope 1 category covers direct emissions from an organisation's facilities. Scope 2 covers emissions from electricity purchased by the organisation. Scope 3 covers other indirect emissions, including those from general suppliers.

The emission estimate covers all the six gases: carbon dioxide (CO_2), methane (CH_4), nitrous oxide (N_2O), and industrial gases: hydrofluorocarbons (HFCs), perfluorocarbons (PFCs), sulfur hexafluoride (SF_6), nitrogen trifluoride (NF_3).

Reduction in SEC directly reduces the CO_2 emission depending upon the carbon content of the fuel being used or the carbon intensity of the electricity sourced from outside. In some cases, energy efficiency measures can indirectly help in reducing the CH_4 emission too. Case in point is bio-gas generation from industrial effluents and utilisation of the same as fuel in the boilers or production of hydrogen.

The energy audit process can be extended to estimate the GHG reduction potential from implementation of the recommended energy efficiency measures. Taking the example of the boiler efficiency improvement case study, we can calculate the GHG reduction as shown in Table 4.4.

TABLE 4.4
Reduction of GHG Emission

Parameters	Units	Values	Data source
Steam generation	TPH	50	DCS log
Steam enthalpy	kcal/kg	750	Steam table
Feed water enthalpy	kcal/kg	120	DCS log
Heat added in boiler	Gcal	31.5	Calculated
Fuel GCV	kcal/kg	2,951	Laboratory report
Thermal efficiency baseline	%	74	Efficiency test
Thermal efficiency project	%	77	Efficiency test
Fuel consumption baseline	TPH	14.4	Calculated
Fuel consumption project	TPH	13.86	Calculated
Carbon in fuel	%	35	Laboratory report
Specific GHG emission	T/T	3.67	Dulong equation
Reduction in GHG	TPH	0.72	Calculated

This way, we can calculate and report on emission reduction on both scope 1 and scope 2 account. However, the accounting methods for scope 3 emission is more complicated. In case of such requirement, it is best to engage professional domain experts at least for audit of the scope 3 emission estimates.

MINIMUM ENERGY APPROACH

Minimum energy approach for energy audit has its origin from the pioneering work done on process synthesis[7,8] synthesis of heat exchanger networking[9] for improving process energy efficiency. These concepts are now applied in many process industries such as Refineries, Petro-chemicals, Chemicals, Fertilisers, Sugar, Paper, and Edible oil. In a very simplistic term, it means heat energy that has not been used in the conversion process is waste by itself. Additionally, more energy will be needed for disposal of the unconverted portion of the energy.

In the case of a steam power plant, the efficiency of the turbine and the condenser determine the heat to power conversion efficiency. If the turbine is not working efficiently or the condenser tubes are dirty, the exhaust steam temperature from the turbine will be higher resulting in wasted heat. We will then have to use more energy for pumping of higher quantity of cooling water in the condenser and more fan energy in the cooling tower for disposing the additional waste heat.

Similarly, we can take the case of different types of process equipment such as evaporators, distillation columns, heat exchangers, where is steam is used for process heating and cold water is used for process cooling. In case these equipment are working inefficiently, say due to fouled up heating surfaces, we will need more steam for the same amount of process load. The extra heat of steam has to be finally disposed in the condenser increasing the cooling load incurring double penalty as in case of turbine described above. By properly networking the heat exchangers and optimising their performance, significant energy savings, sometime to the extent of 30%–50% can be achieved.[1]

The concept of the minimum energy approach can be applied not only to process heating and cooling but also to mechanical and electrical systems. In digital electronic systems, for example, the heat generated by various components needs to be dissipated, often requiring additional power for cooling purposes. By designing more efficient circuits and cooling mechanisms, waste heat generation is reduced. Much energy savings is being realised resulting in lesser power consumptions in recharging of the batteries. The following is an interesting case study on 'minimum energy approach' for reduction of power requirement for mobile phone.

Case Study 4.2: Minimum Energy Need for Mobile Phone

The author was engaged by one of the multilateral development agencies to carry out technical evaluation of few innovative energy efficiency projects seeking funding support. Reducing power consumption in mobile phone was one such project proposed by two young US-returned Chinese electronic engineers. They had carried out an exhaustive study on power consumption by the mobile phones in the market. Their finding was that almost 90% of the current level of power consumption is due to free and unproductive movements of electrons. This power consumption can be significantly reduced by guiding the movement of electrons for productive duty as per requirement of the phone user. They demonstrated a protype phone circuitry developed by them that was consuming about 50% of the current level of power consumption. Interestingly, they also demonstrated how the innovative circuitry developed by them can be retrofitted in the existing mobile phones thereby making it easy to market the product.

This is a perfect case of minimum energy design where the developers were able to identify that only about 10% of the current power consumption was for the productive use, rest being wasted.

Figure 4.10 shows a typical approach for application of MEA concept for energy audit.

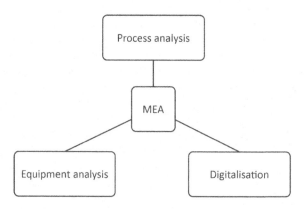

FIGURE 4.10   Minimum energy audit framework.

The objective of this approach is to move away from an equipment-centric approach and instead adopt a systemic approach to energy efficiency.

In the context of water pumping, simply installing an efficient pump may not be enough to ensure overall system efficiency. It is essential to consider the actual demand for water in the process and analyse the hydraulic network to develop a water delivery system that is least energy-intensive. The audit process begins with assessing the water demand for various processes, such as cooling, washing, or flotation. Once the demand assessment is complete, hydraulic network analysis is conducted to identify opportunities for optimising the water delivery system. The current energy consumption of the system is then compared against the energy requirements calculated for the least energy-intensive water delivery system. Field tests are conducted to determine the impact of individual components such as motor efficiency, pump efficiency, and hydraulic efficiency of the distribution system on the identified gap.

Based on the findings of the audit, improvement measures are identified to bridge the gaps and targets are set considering the techno-economic feasibility of the proposed measures. Figure 4.11 illustrates the result of energy audit in a paper mill following the minimum energy approach.

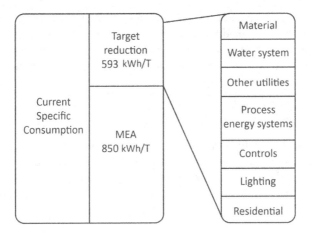

FIGURE 4.11 MEA audit results.

In a traditional audit, we usually aim at 10%–15% reduction in the SEC. MEA audit helps in delayering of the inefficiencies at various levels resulting in identification of measures that reduces the energy consumption by 30% or more in most of the cases. With digitalisation, it will be possible to achieve much higher level of savings.

DIGITALISATION AND ENERGY AUDIT

Most of the industrial organisations today have an overall digitalisation plan woven into their strategic business plan. How does an energy auditor then add value to the digitalisation plan while carrying out energy audit in an industrial plant? And how does it help in improving the quality of audit?

An energy auditor can add value to the digitalisation plan and enhance the quality of the energy audit by focusing on the different components of the digital system:

DATA ACQUISITION

Energy auditors can assess the effectiveness of the data acquisition systems including smart sensors, converters, and transducers. They can evaluate if these components are accurately capturing the required data for energy consumption and process variables. Auditors can identify any gaps in data collection and recommend improvements, such as installing additional sensors or upgrading existing ones. By ensuring comprehensive and accurate data acquisition, auditors contribute to the reliability and validity of energy analysis and optimisation.

ANALYTICS

Energy auditors with knowledge of advanced analytics can review the algorithms, simulation models, and artificial intelligence used in the digital system. They can assess whether these analytical tools consider the impact of different process variables on energy consumption and production output. Auditors can help identify any missing variables or inadequately assessed impacts and suggest modifications to the algorithms to improve energy efficiency. By reviewing and enhancing the analytics component, auditors contribute to optimising energy consumption in the entire production process.

DRIVES

While drives such as actuators, robotics, and 3D printers are primarily focused on process optimisation and automation, energy auditors can assess their responsiveness to the changes in the variables that impact energy efficiency.

DISPLAYS

Energy auditors can also review the display components of the digital system, such as metres, loggers, and monitors. They can assess the functionability of these displays in providing real-time energy consumption and efficiency performance. Auditors can recommend improvements, such as data visualisation techniques or user-friendly interfaces, to enhance the understanding and utilisation of energy-related information by plant operators and management. Highest value from the display system is achieved when operators are able to use the displays for analytics and continuous improvement. Auditors can help in redesigning of the display system for this.

By actively assessing and optimising these components, energy auditors can add value to the digitalisation plan and contribute to improving energy efficiency in industrial plants. Their expertise and analysis can help identify opportunities for energy savings, validate the effectiveness of the digital system and provide recommendations for further enhancements.

EXTERNAL STAKEHOLDERS' ENGAGEMENT – MSME SEGMENT

A separate chapter (Chapter 7) has been devoted on employee engagement for promotion of efficiency and conservation in large organisations. General principles driving the stakeholders' engagement process remains the same for both internal and external stakeholders. However, some unique challenges are always faced in implementing new improvement initiatives in the MSME segment. Success totally depends upon how well we can keep the CEO engaged with the process from beginning to end.

MSME units rarely undertake energy efficiency projects by themselves. Usually, such projects are initiated at cluster levels by governments often with support from multilateral development agencies. Cost of energy audit in individual MSME unit is borne by the project subject to certain conditions such as commitment from the owners on implementation of the feasible energy efficiency measures identified during energy audit. The consultants are appointed by the Government agencies as per terms of engagement that includes linkage of payment to the consultants to success factors in identification of efficiency measures and implementation by the MSME industry. First challenge is faced in attracting the attention of the CEO, next is getting his time and then the most formidable one of negative feedback from any member of his core team of two to three persons. The following case study shows how an innovative engagement strategy was deployed for successful implementation of a project in an MSME unit.

Case Study 4.3: Stakeholders' Engagement: MSME Textile Unit (Export House)

The unit was being directly managed by the Director (Owner) with the help of three managers responsible for purchases, production, and maintenance, respectively. The initial impetus for energy efficiency work came from the Manager (Maintenance) who had attended a seminar on energy audit. With nod from the Director, he appointed an external energy auditor for carrying out a detailed energy audit. The Manager (Maintenance) quit his job during the energy audit process, which led to a change in responsibilities within the unit. This change in personnel can sometimes disrupt the continuity and progress of energy efficiency projects if not properly managed.

The Manager (Purchase), who assumed the additional responsibility of maintenance, displayed resistance to accepting the audit report and scheduling a presentation meeting with the Director (Director A).

The auditors conducted an informal stakeholder's analysis. This analysis revealed that the Director A was not engaged, the Manager (Production) was indifferent, the Manager (Purchase) was a blocker, and the Junior Engineer (Maintenance) was an advocate for energy efficiency.

Leveraging External Influence: To overcome the resistance and indifference of Director A and the Manager (Purchase), the auditors sought help from Director of another unit (Director B) that had implemented energy efficiency projects. An informal meeting was arranged by the Director B. In the meeting, he highlighted Director A's role in initiating many path-breaking initiatives and how improving energy efficiency would enhance their social standing and reputation. This external influence and endorsement had a significant impact on Director A's perception and willingness to implement the audit findings. A presentation meeting was held in Director A's office with participation from other managers and the junior engineer. Director A accepted the audit report and implemented the findings promptly. He then became the ambassador in the cluster for scaling up the programme.

This case study underscores the need for approaching the engagement process with open mind and innovate process based on the principles of 'different stokes for different folks'. It also highlights the need for identifying individual champions within and outside of the organisation who can influence decision-making processes.

CONCLUSION

This chapter has detailed out different types of energy audit and their applications for achieving different objectives on energy performance in industrial facilities. The information provided will come out handy for the energy auditors and the industrial energy professionals in carrying out the audits from a 'seeking solution' perspective. This is in line with the theme of the book that is viewing energy audits as diagnostic studies for delayering the embedded opportunities for improvement.

The significance of conducting high-quality energy audits and the approach for carrying out such audits to reduce energy consumption are highlighted with illustrated case studies. The three common types of energy audits, namely walk-through audits, detailed energy audits, and detailed energy audits with expanded scope and on-site measurements, are explained. The book also demonstrates how integrating analytical concepts like root cause analysis, carbon accounting, minimum energy approach, and digitalisation can greatly enhance the process of discovering energy savings through the audits.

Lastly, the chapter emphasises the importance of a formal strategy for engaging with key stakeholders to ensure successful outcomes from the energy audit. An illustrative case study from an MSME industrial unit is provided to illustrate the approach and its effectiveness.

NOTES

1 Canadian Industry Partnership for Energy Conservation, 2002. https://natural-resources.canada.ca/energy-efficiency/energy-efficiency-for-industry/canadian-industry-program-energy-conservation-cipec/20341
2 Authors database from audit reports of over 20 paper and other mills in India, Kenya, Jordan, Egypt.
3 What is an Energy Performance Indicator (EnPI), 2023, https://50001store.com/articles/energy-performance-indicator-enpi.
4 Tokutaro Suzuki, Overview of TPM in process industries (T&F e-books, 1994). https://www.taylorfrancis.com/chapters/mono/10.1201/9780203735312-1/overview-tpm-process-industries-tokutar%C5%8D-suzuki.
5 Bo Shen, Lynn Price, Hongyou Lu, Energy Audit Practices in China: National and Local Experiences and Issues, *Energy Policy*, 46 (2012): 346–358.
6 Bureau of Energy Efficiency, India (2023). https://beeindia.gov.in/en/programmes/perform-achieve-and-trade-pat.
7 Minimum energy design-Thing, MW, The Chemical Engineer, March 1982.
8 Naonori Nishida et al., A Review of Process Synthesis, *AIChEJ*, 27(3), March 1981: 321–351.
9 Linnhoff Bodo, The efficient use of energy in the process industries-The Chemical Engineer, (October 1980).

5 Implementation of Retrofit Projects

INTRODUCTION

The energy auditors have submitted the audit report. They have identified and listed number of retrofit projects. The projects have been rank ordered for implementation based on projected technical and financial performance and risks, what next?

The facility engineers are normally in a position to implement smaller and simple projects as per provisions in the annual budget on energy savings measures. The larger projects require more detailed analysis taking into consideration the technical and financial risks. External auditors rarely have the capacity to fully comprehend various implementation issues associated with larger retrofit projects. There are different project types that throw up different implementation challenges. A simple project such as replacement of a pump by a more efficient one does not require detailed engineering. But hook up of a retrofit project such as a cogeneration or waste heat recovery system will require more detailed engineering and technical and financial risk analysis. Often the configuration of a project can change post reevaluation of options and comprehensive risk analysis for larger projects. Additional cost may have to be incurred due to these engineering changes. We may also incur revenue loss due to lost production during hook up of the retrofit. Similarly, gains from the proposed project may change either way, positive or negative, due to changes in the energy performance and energy prices. The project proposal has to be therefore, reworked taking into account the recommendations of the energy auditors and the findings from the detailed technical and financial options and risk analysis.

There are several options available to a facility for engineering, procurement, financing, and construction of energy efficiency retrofit project. Appropriate strategy is chosen depending upon the complexity of the project and technical and financial capacity of the project host. Usually, the hosts implement simple, low-investment, and low-risk projects inhouse. Alternative mode of execution such as 'build, own, operate, and transfer (BOOT)' or 'Engineer, procure, construct, and maintain (EPCM)' are preferred choice for complex and high-investment projects such as cogeneration and waste heat recovery. There are specialised agencies such as 'energy services companies (ESCOs)' who undertake implementation including financing of such projects under 'energy-saving performance contract (ESPC)'. Each of these mechanisms has its own merits and demerits. Amongst other benefits, ESPC mechanism usually improve the technical and financial risk profile of the projects at a low incremental cost.

Finally, it is important to carry out evaluation of the technical and financial performance of the projects post implementation. For this, we need to adopt a robust performance measurement and verification (M&V) protocol to assess the actual impact of the project on energy parameters. Considering the critical role of M&V

DOI: 10.1201/9781003415718-5

for evaluation of retrofit projects, a separate chapter has been dedicated to M&V protocol. However, to set the context, the role of M&V for successful implementation of energy-saving projects have been deliberated in brief in this chapter too. Digitalisation plays a very important role in enhancing performance of energy efficiency retrofit projects in varieties of ways. More interactive control algorithms help in optimisation of processes. Similarly, real-time tracking of the performance enhances the sustainability of energy savings. In fact, digitalisation can automate the M&V process thereby eliminating need for human intervention and making the evaluation system transparent and robust.

This chapter has been accordingly designed to comprehensively address the implementation issues discussed above backed up with case studies on actual implementation. The chapter has been organised as under.

Preparation of projects
Risk analysis
Project engineering, costing, and financial analysis
Mode of implementation

A significant portion of the chapter has been devoted to discussing case studies related to specific implementation issues that have been previously highlighted. These case studies provide real-world examples and practical insights into the core of the subject matter being discussed.

PREPARATION OF PROJECTS

The energy audit reports generally classify the identified energy efficiency measures under the following three categories:

Short-term measures to be implemented on priority, simple projects with high
    financial return
Medium-term projects to be taken up as per capital budget of a facility
Long-term projects, usually larger investment with longer gestation

The projects under each of these three categories can further be classified into project types highlighting the implementation issues as presented in Table 5.1.

It is seldom that all the projects listed in the audit reports are taken up for implementation. Investment attractiveness, both size of investment and return on investment; complexity of the project and measurability of the energy and financial savings are the important criteria for selection of the project for implementation. Lighting retrofit projects are the simplest to implement, more so when the project is configured for like-to-like replacement. The projects around industrial motor systems usually offer attractive option. Multiple technology choices are there for achieving the same energy efficiency goal, few such examples include:

Replacement of the equipment by more efficient ones
Proper sizing of equipment

TABLE 5.1
Project Types and Implementation Issues

Project Type	Illustrative Examples	Implementation Issues
Operation management	Maintaining pressure and temperature profiles of steam for processes	Management of information, automation
Repair and maintenance	Poor insulation, choked pipelines, scaled heat exchangers	Management of information, quality of investigative studies/energy audit
Replacing inefficient equipment	Lights, pumps, fans, compressors	Proper financial analysis-cost and savings, quality of energy audit
Improving system efficiency	Revamping hydraulic network, fan ducting, compressed air system	Technical risk analysis, quality of energy audit
In process heat recovery	Heat exchanger networking, high-efficiency heat exchangers	Quality of engineering analysis, technical and financial risk analysis
Cogeneration	Fertilisers, chlor-alkali, paper mills, sugar mills	Quality of engineering analysis, technical and financial risk analysis, project financing
Waste heat recovery	Iron and steel plants, cement plants	Quality of engineering analysis, technical and financial risk analysis, project financing

Trimming of impellers
Installation of variable speed drives (VSD), there are alternative technologies for VSD too
Proper sizing and engineering of pipelines/ducts
Real-time efficiency monitoring through digitalisation

Even though these projects are simple, there could be multiple technology options for achieving higher efficiency as well as meeting the feasibility criteria. Case Study 5.1 demonstrates the methodology that was adopted for evaluation of different options before freezing the option for implementation.

Case Study 5.1: Technical Options and Their Evaluation for Improving System Efficiency for Process Water Pumps

This case pertains to implementation of energy efficiency improvement measures for few process water pumps in a paper mill under ESPC. The energy audit identified the improvement measures by undertaking efficiency tests for individual pumps. Energy savings summary report was prepared based on the set efficiency targets, estimated energy and financial savings and normative capital cost, as shown in Table 5.2.

TABLE 5.2

Process Water Pumping Retrofit Opportunities

	Power kW		Efficiency			Savings Potential		Capex	Payback
Pump duty	Hydraulic	Input	Current	Target	kW	kWH/year	$/year	$	Months
Reservoir	44	110	40%	75%	39	323,400	20,913	12,000	6.9
Brown stock washer	57.5	142	40%	75%	49	411,600	26,617	25,000	11.3
Unbleached centricleaner	32	84	38%	70%	27	225,120	14,558	5,000	4.1
Bleached centricleaner	34	82	41%	75%	28	231,000	14,938	5,000	4.0

REVALIDATION OF AUDIT FINDINGS

More detailed study was carried out for the revalidation of the audit findings and identification of energy losses in the individual components in each of the pumping system. Operating data were collected afresh over extended period through field measurements, log books, and historical records. The field measurements were done with calibrated meters recording readings every 10 min for duration of 2 h. Three sets of readings were taken at different times to observe the change in system behaviour with changes in the operating conditions such as loads and fluctuation in the power supply system. The operating efficiencies of the individual pumps were calculated taking the motor efficiency from the nameplate and calculating the throttling losses based on head loss across the throttle valves. Sankey diagrams were developed for all the pumps under study, an illustrative example for one of the reservoir pumps is shown in Figure 5.1.

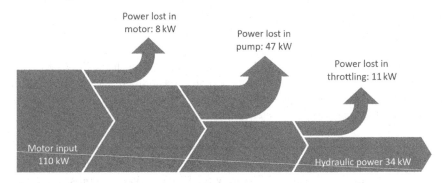

FIGURE 5.1   Sankey Diagram – Reservoir pump.

Energy-saving potential by improving efficiency were estimated for all the operating pumps following the same methodology. The results are tabulated in Table 5.3.

All the pumps were inefficient except that the extent of inefficiency of the brown stock washer pump could not be determined. It was not possible to measure the

TABLE 5.3
Break up of Losses

| Duty | Break up of Losses kW (%) | | | |
	Motor	Drive	Pump	Throttling
Reservoir	8 (12%)	0	47 (71%)	11 (17%)
Brown stock washer	15 (14%)	51(49%)	a	38 (36%)
Unbleached centricleaner	8 (10%)	0	43 (90%)	0
Bleached centricleaner	8 (10%)	0	40 (90%)	0

ᵃ Included in the drive loss figure taking the input power as total power minus loss in motor

power loss in the gear drive. There was also significant power loss in throttling for the brown stock washer pump. Replacement of all the pumps and providing VFDs for reduction of the throttling losses in the brown stock washer pumps were considered for implementation. It was also decided to carry out further analysis to assess the feasibility of retrofitting of new pump alone for the brown stock washer pump considering the mechanical constraints.

CONSTRAINT ANALYSIS–BROWN STOCK WASHER PUMP RETROFIT

Brown stock washer pumps were supplied by the vendor as an integral part of the washing systems. The foundation and connectivity with the driven shaft of the drive unit and water connection coordinates were also not matching with high-efficiency pumps available in the market. Modifications of the foundations and other retrofits will require shutdown of the system, which will cause production loss. As such, it was decided to drop the measure except for retrofit of a variable frequency drive (VFD) for reduction of throttling losses.

PROJECT LIST FOR IMPLEMENTATION

It was decided to replace pumps for the all the applications except for the Brown stock washer pumps and VFD for the later.

The above case study shows a typical approach for evaluation of options before finally selecting an option. Some projects such as a cogeneration or a waste heat recovery (WHR) requires more detailed analysis for ascertaining techno-economic feasibility. Such projects face M&V challenges too due to difficulty in measuring some of the input parameters. We have to therefore, carry out further technical and financial risk analysis for such projects as presented in the next section.

RISK ANALYSIS

Retrofit energy efficiency projects face engineering, M&V, and financial risks. In this section, we will address the technical and financial risks leaving the M&V risks, for which a separate chapter has been devoted. Conducting a comprehensive risk analysis is a crucial step for high-investment projects. Such analyses help project stakeholders identify, assess, and mitigate potential risks. Few of the key considerations for a comprehensive risk analysis include technology, financial, market dynamics, environment, and sustainability and finally organisation capacity.

In case of new and innovative technologies one has to assess the risks associated with its development, implementation, and operation and maintenance. Financial risk arises from potential cost overruns and/or revenue shortfalls. MSME industries are particularly susceptible to market risks due to potential change in the market that could impact the demand for the products. Efficiency projects do not generally face regulatory or environmental risks. However, there are exceptional cases where cost-based approach is adopted by regulators for approval of tariff. In such cases, gains from efficiency can get neutralised due to pass through nature of the cost. High-technology projects often fail due to lack of organisation capacity to absorb such technologies. In such cases, it is always desirable outsource the operation and maintenance till such time internal capacity has been created.

Table 5.4 illustrates a risk assessment model for energy efficiency that can be used for quantification of risks and development of a risk assessment score card.[1]

Technical risks for simple and low-investment retrofit projects are low. This is so as the technologies are fairly mature and one does not have to do engineering for configuration

TABLE 5.4
Comprehensive Risk Analysis Matrix

Phase	Risk Indicator	Risk Scorecard (Potential Impact)
Project development	Quality of energy audit-quality of projects	High
	Determination of energy use baseline	Medium
	Baseline adjustment methodologies	Medium
	Assessment of financial cost and benefits	High
	Cash flow assessment	High
Implementation	Complexity of measures	High
	Reliability of technology	Low
	Capacity of the implementation contractor	Medium
	Quality of equipment	High
	Quality of contract	Low
Operation	Capacity of the operator	High
	Level of end-user participation	Low
	Previous O&M experience	Medium
	Product warranties and guarantee	Medium
	Availability of technology maintenance support services	High
	Quality of O&M contract	Medium
	Verification of saved energy (PMV)	Medium
	Cash flow management	High

TABLE 5.5
Project Technical Risks

Technical Risks	
Low	**High**
Energy efficient motor systems, modification/ retrofits in existing equipment such as boilers and furnaces	Waste heat and top pressure recovery systems
	Cogeneration
Variable speed drives	Integrated energy efficiency and renewable energy
Process control and automation	Integrated supply and demand side management
Lighting and HVAC	Advanced process optimisation technologies
Data analytics and remote monitoring	

of such projects. However, risk increases for projects, such as WHR and cogeneration projects requiring more detailed engineering for the project itself and hooking up with the existing system. Table 5.5 shows a list of typical low- and high-risk projects.

Energy auditors carry out risk assessment study for projects in both the categories based on the prevailing situation in a particular facility. It may not be worthwhile to incur cost in repeating such a study afresh at the time of investment decision for a low-investment low-risk project. However, for high-investment and high-risk projects, it is always prudent to carry out such a study afresh and in more detail before committing investment.

The following case study on installation of a Topping turbo-generator (TG) project shows how the configuration of the project had to be changed as a result of risk analysis post selection of the project for installation. This project was characterised by many of the risks outlined in the table on comprehensive risk matrix (Table 5.4).

Case Study 5.2: Risk Analysis and
Mitigation-Topping Cogeneration Project

The project: Installation of a Topping TG set for generation of more power (exergy potential) by improving cycle efficiency in a captive cogeneration plant in a chemical plant.

The existing supply side energy system (Figure 5.2) consisted of:

Four boilers each of 50 TPH steam capacity at 35 bar pressure and 450°C temperature, one double extraction condensing TG set of 10 MW capacity operating at 33 bar pressure, 445°C temperature, one straight condensing TG set of 10 MW capacity operating at 15 bar pressure, 375°C temperature and one pressure reducing cum desuperheating station of 80 TPH capacity at 35/15 bar pressure and 450/380°C temperature.

About 175 TPH steam generated in the four boilers at 35 bar g pressure and 450°C temperature feed a common high-pressure steam header. Of this, 75 TPH steam is fed into the 10 MW high-pressure double extraction and condensing TG set for meeting part of power and process steam demand at 8.5 and 2.5 bar g, respectively.

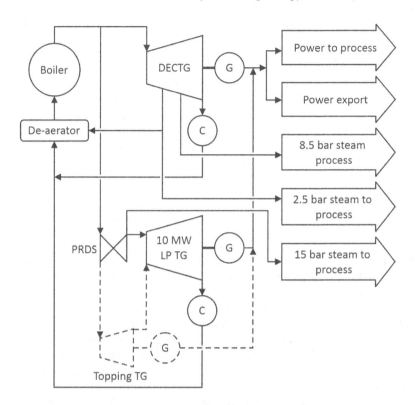

FIGURE 5.2   Topping turbine case study. DEC TG, double extraction condensing turbo-generator; PRDS, pressure reducing cum desuperheating station; LP, low pressure.

Another 80 TPH steam is passed through a pressure cum desuperheating station (PRDS) for supplying 60 TPH steam as inlet to the low pressure 10 MW TG set and 20 TPH steam for process.

The project consisted of installing a back pressure TG set for reducing the pressure from 33 to 15 bar g, thereby eliminating the PRDS. The power potential is calculated using the thermodynamic table:

Determination of the isentropic heat drop from 33 to 15 bar g system

Calculating the power potential as per the following equation:

$$P = Q_s \times h_i \times \eta_e \times \eta_m / 3,600$$

where P, power output in MW; Q_s, flow of steam in TPH; h_i, isentropic heat drop from 33 bar g 445°C to 15 bar g; η_e, isentropic efficiency of expansion in the turbine; η_m, mechanical and electrical efficiency of conversion.

The power potential has accordingly been calculated as shown in Table 5.6.

The project was financially very attractive as will be seen from Table 5.7.

The auditors carried out a risk assessment study. Table 5.8 summarises their observation and recommendation.

TABLE 5.6
Topping TG Set Capacity

Isentropic Drop from 33 to 15	Units	33 Bar g System	15 Bar g System
Pressure	bar g	33	15
Temperature	°C	445	332
Enthalpy	kJ/kg	3,327	3,106
Flow	TPH	80	80
Entropy	J/kg K	7,006	7,007
Isentropic drop	kJ/kg		221
Isentropic efficiency (η_e)	Factor		0.65
Electro-mechanical efficiency (η_m)			0.95
Actual drop	kJ/kg		143.7
Actual enthalpy	kJ/kg		3,183.4
Actual temperature	°C		363
Actual entropy	J/kg K		7116
Power output	MW		3.03

TABLE 5.7
Financials of Topping TG Project

Power gen	MW	0	3.03
Steam use for power	TPH		3.45
Specific steam consumption-project	T/MWH		1.14
SSC-existing	T/MWH		5.5
Savings	T/MWH		4.36
Steam cost	US$/T		25.00
Load factor	Factor		0.8
Annual operating	Hours		7000
Annual savings	US$		61,0539
Capex	US$		1,150,000
ROI	%		53%
Payback	Years		1.88

Accordingly, auditors recommended implementation of the project subject to selection of a reputed vendor having experience in implementation of such projects.

The project was presented to the board of the company for approval. The projected financial return was highly attractive. Moreover, early execution of the project

TABLE 5.8
Risk Assessment Topping TG Project

Phase	Risk	Risk Scorecard
Project development	Quality of energy audit-quality of projects	Low. Calculation of energy savings based on thermodynamic principles
	Baseline, baseline adjustment	Nil
	Financial and cashflow	Low as no externalities are involved besides attractive ROI
Implementation	Complexity of measures	High, reputed vendor and turn-key implementation arrangement
	Reliability of technology	Low
Operation	Capacity of the operator	Low as the plant is already operating a cogeneration plant
	Verification of saved energy (PMV)	Nil

would help in meeting the anticipated increase in demand for power due to expansion of the chemical plant. However, the project was sent back for further technical and financial analyses considering the mitigation of the following risks:

- Engineering of the project considering the complexity, as this would be a first of its kind
- Availability of a reputed vendor for the Topping TG set as well as control system
- Any instability would impact not only the power plant but also the operation and safety of the chemical plant

It was decided to carry out a risk analysis afresh engaging experts conversant with Topping TG technology taking the following steps:

- Reassess the project capacity
- System stability and reliability – 'What-if' analysis
- Re-engineering for risk mitigation
- Costing and financial analysis

RISK ASSESSMENT AND MITIGATION

Outlines various risks as perceived by the host facility are as follows. These risks were deliberated in detail in several brainstorming sessions with participation of engineers from the host, energy audit team and external experts. We identified two major technical risks that require mitigation by proper engineering of the project. These are summarised as follows.

Risk 1

Compatibility of the hydro-mechanical governing system of the existing 10 MW TG set with that of the proposed turbine.

Mitigation 1

Replacing the governor of the 10 MW TG set with electronic governor along with electro-hydraulic actuators.

Risk 2

Impact of tripping of the entire 13 MW system in the event of tripping/failure of the Topping turbine on the operability and safety of the chemical plant.

Mitigation 2

Electrical load management for disconnecting load equivalent to supply from the Topping TG set and isolating the 10 MW set from the impact of disturbance in the Topping turbine.

For this, provision was made for a separate electrical panel supplying power to non-critical loads. The incomer circuit breaker panel was linked to the master relay of the Generator master relay of the *Topping TG set.*

Mitigation 3

Design and development of a tracking cum actuating system for maintaining steam supply to the 15 bar g system without interruption. The system (Figure 5.3) consisted of:

1. A controller tracking the position of the inlet steam control valves of the Topping turbine
2. A 33/15 PRDS station following the tracking controller
3. A quick acting hydraulic isolating cum non-return valve before the PRDS
4. A 50 mm bypass valve across the quick closing valve to keep the line always heated up

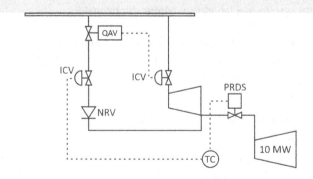

QAV: Quick acting valve, **NRV:** Non-return valve, **PRDS:** Pressure reducing cum de-superheating station, **TC:** Temperature controller, **ICV:** Inlet control valve

FIGURE 5.3   Schematic control system.

The tracking controller was maintaining the opening position of the PRDS to allow the same amount of steam flow as was from the Topping turbine before the disturbance. The hydraulically operated quick acting isolating cum non-return valve was connected to the safety hydraulic system of the Topping turbine. In the event of tripping of the Topping turbine, the hydraulic pressure that was keeping the quick acting valve closed will fall thereby opening the quick acting valve even before tripping of the control valves of the Topping turbine. Steam supply will therefore, be maintained in the 15-bar g system without interruption as the PRDS was already open as per the signal from the tracking controller.

A simulation model was developed with a view to assess the stability of the proposed control system. This was tested in the control system laboratory of a reputed Germany-based TG vendor. The result was found satisfactory

The cost of the project was revised accounting for the cost of the following additional items:

1. Electronic governor along with sensors and actuators for the existing 10 MW TG set
2. Electrical panel along with cables for shifting some of the non-critical loads from the existing 11 kV panels to the new panel
3. An integrated PLC-based control system along with actuators and sensors for tracking the control system of the Topping TG set and bringing in the 33/15 PRDS station in case of disturbance/tripping of the Topping TG set
4. A quick acting isolating cum non-return valve in the 33 bar g pipeline to the 33/15 PRDS operated by the safety oil system of the Topping TG set
5. Design, engineering and development of the simulation system for testing out the entire control system in a laboratory before shipment

The additional cost amounted to almost equal to the original cost estimate for the project. Even then, the project was financially quite attractive, as shown in Table 5.9.

TABLE 5.9
Revised Financials: Topping TG Project

Particulars	Units	Value
Power generation Topping TG	MW	3.03
Steam consumed for power in Topping TG	TPH	3.45
Specific steam consumption	T/MWH	1.14
Specific steam consumption-baseline	T/MWH	5.5
Savings in specific steam consumption	T/MWH	4.36
Steam cost	US$/T	25.00

(Continued)

TABLE 5.9 (*Continued*)
Revised Financials: Topping TG Project

Particulars	Units	Value
Load factor	Factor	0.8
Annual operating	Hours/year	7,000
Annual savings	US$/year	610,539
Financial Projections at Proposal Stage		
Capital cost (Capex)	US$	1,150,000
Simple ROI	%	53%
Payback period	Years	1.88
Revised Financial Projections		
Capital cost (Capex)	US$	1,800,000
Simple ROI	%	34%
Payback period	Years	2.95

Explanatory notes: Table 5.9 on financials

Steam consumed for power generation in the Topping TG set is calculated as per following equation:

$$Q_{st} = \frac{h_{ac}}{h_s} \times Q_s$$

where Q_{st}, Steam consumed in the Topping TG set; h_{ac}, Actual heat drop in the Topping TG set; h_s, Enthalpy of inlet steam to the Topping TG set; Q_s, Steam flow through the Topping TG set;

Steam cost at US$25/T taken from the host internal cost sheet at the time of appraisal of the project

This case study shows how the configuration of a retrofit project changed as a result of more elaborate risk analysis. In this particular case, the project cost increased significantly. However, the project was still very attractive financially. The project was implemented. The project generated better financial results as a result of increase in price of the input fuel.

The above-described project did not have to face market risk as there was demand for additional power from the mother chemical plant. Market risk is faced by such projects in case of sale of power to the utilities. Power generation from waste heat deploying Rankine cycle had been adopted by the direct sponge iron-making plants in many countries. Similarly, large number of sugar-manufacturing plants across the globe have made large investment in advance bagasse-based cogeneration technologies for generating additional revenue from sale of surplus power. The viability of investment in such projects depend upon offtake of surplus power by the Power Utility companies at the agreed tariff. Utilities purchase power from these systems as per long-term contract, called power purchase agreement, mostly for meeting their renewable energy offtake obligations. Reduction in cost of renewable alternatives, such as solar and wind energy, made it financially more attractive for utilities to shift to these alternatives. In such cases, market assessment study becomes more important than technology risk assessment.

PROJECT ENGINEERING, COSTING AND FINANCIAL ANALYSIS

Energy efficiency measures identified during the process of audit are presented as replacement or retrofit options depending upon the feasibility. We can take the example of a water pump project. The auditors carry out option analysis considering trimming of the impeller or replacement of the pump. The cost of trimming is insignificant but energy saving is also much lower compared to replacement option. The auditors usually carry out the cost benefit analysis by obtaining budgetary quotes for the replacement pump from the vendor and energy-saving estimate. At the time of installation, one may find that the dimension of the new pump does not match with the one being replaced. The cost of the project in such a situation can increase significantly for various engineering modifications for the retrofit.

For a simple like-to-like replacement job, the indicated cost in the audit report can be taken as the final cost. However, more detailed engineering is required for other kinds of retrofit to take care of mis-match in foundation, duct/nozzle orientation and size, electrical cable size and entry, availability of appropriate panel in the power control centres, etc. At a minimum level, we need to prepare the following drawings for freezing the project boundary, procurement specifications, and quantification of required hardware and construction items.

TABLE 5.10
List of Drawings

Drawings	Required for
Civil engineering	Space for housing of retrofit equipment and anchoring
Layout	Construction/modification of foundations
Foundation	Laying of pipelines, electrical and communication cables
Trenches/Racks	Estimation of construction quantity and cost
	Procurement of construction materials
Mechanical engineering	Interconnectivity of retrofit equipment with existing
Piping and instrumentation (P&ID)	equipment
Piping layout and isometric	Preparing schedules (bill of materials and quantities,
Ducting layout and isometric	BOM and BOQ))
Structures	Major equipment
	Pipes and ducts
	Structures
	Field and control instruments
	Control algorithms

(Continued)

TABLE 5.10 (*Continued*)
List of Drawings

Drawings	Required for
	Piping stress analysis and supports
	Compressed air requirement for instruments
	Fire protection system
Electrical engineering	Interconnectivity with existing system
Single line diagram	Review of fault capacity and system safety
Protection schematics	Protection co-ordination
Earthing	Schedule (BOM and BOQ)
	Major equipment
	Switchgears
	Protective relays
	Earthing materials
	Panels
	Cable and busbars, terminations
Information and communication technology	DCS interconnectivity
Information hierarchy chart	M&V reports; real time, historic
Data archival and retrieval flow charts	Management information system (MIS); real time,
Logic diagrams	historic

PROJECT COSTING

The cost of a project consists of:

> EPC cost
> Pre-operative costs
>   Engineering and consultancy
>   Project office including manpower
> Financing cost
> Post-operative cost-M&V

Some of the costs are based on quotes from vendors and service providers, whereas there are few that are assumed on the basis of norms or historical data.

In case of another, the terms of the energy audit included converting identified energy efficiency measures into executable projects. The steps taken for working out the project specifications have already been deliberated and included in the case study presentation. Table 5.11 shows how the project cost was then determined for this project taking into account the individual components and their cost.

The EPC cost of the project has increased from US$ 3.71 million taken at the stage of audit to about US$ 4.1 million after adding for EPC costs resulting in reduction of the return on investment (ROI) from 38% to 35%.

In case of long gestation projects, the cost may increase further adding for pre-operative (primarily cost of the project office), financing, and M&V cost.

TABLE 5.11
Project Costing

Sl. No.	Particulars	Basic Cost (Actual) '000US$	Freight (Normative) 2.50%	Taxes (Normative) 2.00%	Insurance (Actual) 0.03%	Total '000US$
A	Major equipment					
1	FD fans (New fans plus motors)	420	10.5	8.4	0.126	439.0
2	PA fans (VFD)	830	20.75	16.6	0.249	867.6
3	ID fans (VFD)	1,280	32	25.6	0.384	1,338.0
4	ID fans (VFD plus high-efficiency fans)	680	17	13.6	0.204	710.8
5	Boiler feed pump (New pump)	500	12.5	10	0.15	522.7
	Total (A)	3,710	92.75	74.2	1.11	3,878.06
B	Hooking up items	Base cost	Freight 5%	Taxes 3%	Insurance 1%	Total
1	Pipeline	5.0	0.3	0.1	0.1	5.4
2	Valves and fittings	5.0	0.3	0.1	0.1	5.4
3	Ducts	2.5	0.1	0.1	0.0	2.7
4	Cables	5.0	0.3	0.1	0.1	5.4
5	Earth strips	2.0	0.1	0.1	0.0	2.2
6	Consumables	1.5	0.1	0.0	0.0	1.6
	Total (B)	21.0	1.1	0.5	0.2	22.8
C	Total cost of procurement	3,731.0	93.8	74.7	1.3	3,900.85
D	Construction	Normative @ 5%				195.04
E	Total EPC cost					4,095.89

PROJECT FINANCIALS

The financial analyses are carried out at two levels. The snap shot analysis on simple ROI and payback period at the time of audit and more rigorous analysis capturing the financial performance over the life of the project capturing time value of the cash flow.

Table 5.12 shows the snapshot of project financial.

TABLE 5.12
Snapshot-Project Financials

Savings Mn US$/year	Investment Mn US$	ROI %	Payback Months
1.43	4.10	35%	2.87

This snapshot analysis suffices for smaller projects with low gestation and equity financing by the host company.

For larger projects, we determine the net present value (NPV) and the internal rate of return (IRR) capturing future cash flows over the investment's lifetime, discounted to the present value. This is determined without considering the impact of external factors such as inflation on the cashflow. The return so calculated is therefore, called IRR. The discounting factor is usually an assumed figure based on the interest rate for loan capital and expected return on equity by the host company. The following tables present the financial assumptions (Table 5.13).

TABLE 5.13
Financial Assumptions

Particulars	Units	Values
Project cost	'000 US$	4,096
Debt: Equity	Ratio	1
Interest on debt (loan)	%	5
Implementation period post first loan disbursement	Months	6
Moratorium period	Months	6
Project savings	'000 US$/year	1,430
Loan amount	000 US$	2,048
Loan tenure	Months	90
Repayment installments	Months	84
Repayment	'000 US$/month	24.38
Tax rate on profit	%	25
Discount factor	%	6

These assumptions are very specific for a particular facility. In case more time is required for negotiations for loan, equity funding may offer a better option. This helps in early commissioning of the project and deriving financial savings without loss of time. The cost of equity on the other hand is usually higher than the cost of loan.

The financial ratios are then calculated based on these assumptions as shown in Table 5.14.

TABLE 5.14
Project Financials

Particulars	Units	Year 1	Year 2	Year 3	Year 4	Year 5	Year 6	Year 7
Operating	Months/year	12	12	12	12	12	12	12
Savings	'000US$/year	1,430	1,430	1,430	1,430	1,430	1,430	1,430
Less loan interest	'000US$/year	75	88	73	58	44	29	15
Less depreciation	'000US$/year	614	522	444	377	321	273	232

(*Continued*)

TABLE 5.14 (*Continued*)
Project Financials

Particulars	Units	Year 1	Year 2	Year 3	Year 4	Year 5	Year 6	Year 7
Profit before tax	'000US$/year	741	820	913	994	1,065	1,128	1,184
Tax	'000US$/year	127	173	308	333	347	352	356
Net profit after tax	'000US$/year	613	647	604	655	718	776	827
Net Cash accruals	'000US$/year	1,227	1,169	1,048	1,033	1,038	1,048	1,059
Repayment of loan	'000US$/year	146	293	293	293	293	293	293
DSCR	Ratio	8.39	4.00	3.58	3.53	3.55	3.58	3.62
IRR	%				21			

Normally, finance professionals carry out the financial and viability analysis based on the hard cost of investment and savings data provided by engineers. It is possible for engineers to carry out financial analysis using standard spreadsheet tools. By having a good grasp of project financials, engineers can better understand the economic viability and potential ROI of their projects. Here are some key parameters that engineers should consider when analysing project financials:

Initial Investment: This refers to the upfront costs required to implement the project. There are hard costs such as equipment, labour, and other construction costs and soft costs such as project preparatory cost including apportioned cost of audit, royalties, cost of finance, taxes.

Operating and Maintenance Costs: This includes expenses for energy, maintenance, repairs, and replacements.

Energy Savings: Engineers need to work closely with finance professionals to establish the baseline energy cost and cost post implementation of the project.

Project Lifetime: Establish rationale for making assumption of the project life based on scenario analysis

Discount Rate: The discount rate is used to calculate the present value of future cash flows. It reflects the time value of money and helps assess the project's NPV or IRR.

Payback Period: This parameter indicates the time required for the project's savings to recover the initial investment.

By understanding and incorporating these key parameters, engineers can collaborate effectively with finance professionals and ensure that project financials are properly evaluated. It enables them to make informed decisions and prioritise projects based on their economic feasibility and potential benefits.

It is also important for engineers to understand the significance of the key parameters and the impact of project cost and savings. There are competing projects in every organisation seeking funds. The first screening of such projects is usually based on IRR criteria. Most of the organisations have a benchmark IRR for looking at the investment proposals. The projects qualifying through the benchmark filters are then are rank ordered on the basis of projected IRR.

The external funding agencies including commercial banks rate projects based on the debt service cover ratio (DSCR). DSCR is the ratio of cash accrual from the project in a particular year to the loan repayment obligation for the year. The DSCR value indicates the likely risk of a project to default on repayment obligation. Higher the DSCR, more is the capacity of the project to meet this obligation.

The cashflow towards tax payment adversely impacts both the DSCR and IRR values. One can seek tax exemptions from Government for EE projects with low IRR and DSCR if such projects otherwise make significant contribution in reducing the GHG emission.

MODE OF IMPLEMENTATION

The project cycle for an energy efficiency retrofit begins with the energy audit and

Audit	Engineering	Financing	Procurement	Construction	M&V
• Internal • External • Mixed	• Internal • External consulting • Joint review	• Equity • Equity & debt • 3rd party	• Internal	• Internal • External • Mixed	• Internal • Joint • 3rd party

FIGURE 5.4 Steps for implementation of energy efficiency projects.

ends with verification of actual energy savings. Various other steps in the project cycle are shown in Figure 5.4.

Energy-saving measures identified during audit are converted to projects for feasibility analysis and implementation. For some large-scale projects such as a WHR, the process begins with feasibility study as energy audit is usually not required for configuration of such projects.

Depending upon the technical and financing capacity, a facility takes a decision on the implementation mode. The facility can carry out all the tasks internally or outsource some or all of these tasks. There are pros and cons of these two different modes of implementation. We have discussed about the skill competency required for undertaking different types of energy audits. Creating an internal organisation for energy audit and implementation of projects internally enhances the capacity for undertaking all kinds of retrofit projects. However, creating such a team requires deployment of higher quality human resources, their training and capacity building for undertaking such tasks. Some additional capital cost is also incurred for procurement and maintenance of hardware and software for conducting audits. The quality of audit by an external auditor is likely to be better. They bring in knowledge gathered from audits of facilities across the globe in different sectors. They also have better access to information on the implementation issues. By addressing these issues at the early stage, it becomes easier to overcome implementation barrier.

The projects discovered by an internal team on the other hand have better buy-in from the operation and maintenance personnel. This facilitates faster implementation. An internal audit team finds it easier to impart training-classroom, and on the job, for sustaining the energy performance.

Based on the authors' own experience as a facility engineer in the early stages and later as consultants, it has been observed that a hybrid system best suits larger industries. The hybrid system consists of:

Engaging external agencies periodically, say once in 5 years to carry out high-quality audit such as target-driven investment grade audit, minimum energy design.

Creating an internal team on rostering basis to carry out annual audit. The rostering team members, trained on energy audit, is brought together for a limited period to carry out the audit. The team is disbanded on completion of the audit and the members rejoin their normal operation and maintenance duty stations.

For smaller facilities, it is best to engage external agencies certified to carry out energy audit.

Engineering for smaller retrofit projects, such as replacement of inefficient equipment is best done internally, provided there is internal capacity to carry out the engineering tasks. However, for complex projects, such as cogeneration or WHR, engaging experienced external agencies is a better option.

Energy efficiency projects can be financed utilising traditional equity and debt finance, often available at concessional rates from commercial banks. Energy efficiency projects make unique contribution in reducing GHG emission, direct reduction due to lower energy consumption and indirect contribution due to equivalent reduction of energy generation from fossil fuel (IEA).[2] Various global development agencies such as World Bank, International Finance Corporation many other such agencies from all the continents offer different types of financial products for energy efficiency projects. It is possible to further monetise the energy savings by participating in the local and global carbon markets.

One issue, which is drawing lot of attention is the role of energy management system in reducing energy consumption. Best results are achieved when standard protocols such as ISO 50001 energy management standards are followed for managing the entire project cycle. This not only improves the quality of the project but also ensures sustainability of the project performance over its life. US Department of Energy (DOE) has a scheme for recognising the industries that have achieved superior energy performance by instituting ISO 50001 system. An independent third party conducts an audit to verify achievements. Entities that exceed certification requirements can use the Scorecard to achieve DOE recognition at the Silver, Gold, or Platinum level. [3]

There are expert agencies, who provide turn-key services under performance contract spanning the entire project cycle from energy audit to implementation of projects and M&V of savings. Such services are provided under different business models described hereinunder.

ESPC by ESCO
Guaranteed energy savings
Shared financial savings
   Outsourcing
Supply side performance contract
Chauffage-integrated supply and demand side performance contract
Utility DSM programme

ESPC and chauffage services generally cover all the market segments: industrial, commercial, residential, while utility DSM programmes generally cover building segments in both residential and commercial sectors.

ENERGY-SAVING PERFORMANCE CONTRACT

ESPC is a contractual arrangement between a facility and an ESCO for identification and implementation of energy upgrade projects with or without financing by the ESCO. In case of financing by ESCO, the contract follows the shared savings model, while it is guaranteed savings model when financed by the facility. Figures 5.5 and 5.6 illustrate these two models.

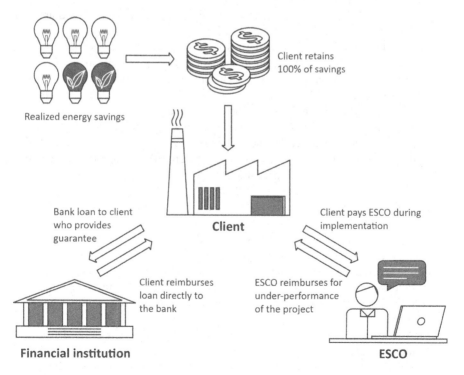

Realized energy savings

Client retains
100% of savings

Bank loan to client
who provides
guarantee

Client

Client pays ESCO during
implementation

Client reimburses
loan directly to
the bank

ESCO reimburses for
under-performance
of the project

Financial institution

ESCO

FIGURE 5.5   ESCO ESPC-guaranteed savings model.

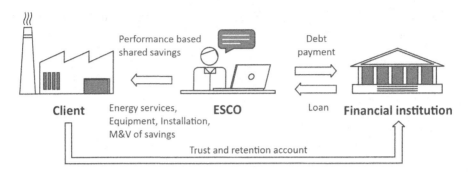

FIGURE 5.6   ESCO ESPC-shared savings model.

In the guaranteed saving model illustrated in Figure 5.5, ESCO is responsible for all the technical aspects of the project and bear the project technical performance risks fully. The host facility pays ESCO for the services plus a bonus based on the performance. In case of shortfall in energy savings, ESCO remains obligated to make up for the shortfall. Host facility finances the project either on its own or taking loan from banks. Host facility bears the financial risk.

Under shared savings model illustrated in Figure 5.6, ESCO remains responsible for financing of the project in addition to the engineering, procurement, construction and operation services during the tenure of the project. ESCO therefore, bears both the technical and financial risks.

The key differences between a traditional service contract and an ESPC are:

The revenue flow to an ESCO under ESPC is 100% linked to the actual energy savings. In a conventional contract, only a small component (about 10%) of the payment is usually linked to performance of the service provider. This means that bulk of the financial risk is shared by the project participants as per a risk allocation matrix developed and inbuilt in the contract.

In addition to the performance risk, ESCOs can face cashflow risk if the facility fails to transfer the cash generated from energy savings to the ESCO. For this, a cash flow security mechanism is developed and maintained during the contract period as outlined in Figure 5.7.

The key features of the arrangements are:

A trust and retention account (TRA, also called Escrow) is created for managing the cash flow during the project tenure

ESCO is primarily responsible for generating adequate cash flow from energy savings to meet all payment obligations and makes up for any shortfall

Host facility is responsible for transferring the cash flow from energy savings to the TRA

TRA distributes the cash to individual stakeholder as per the defined 'water fall structure'

At the end of the project tenure, TRA is extinguished and all the rights are transferred to the host facility.

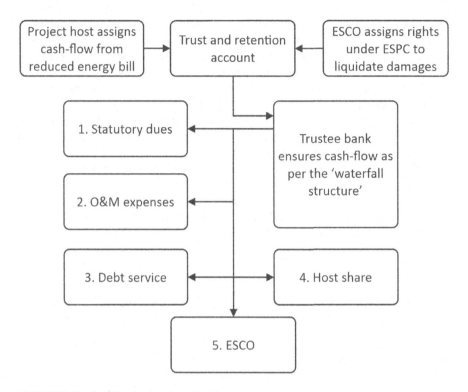

FIGURE 5.7   Cashflow under shared savings contract.

ESPC by ESCOs are successfully operating in many countries. China has been the most successful country in promoting ESCOs and ESPCs in the industrial sector. In 2020, China's total investment in ESCO projects grew 12.3% to USD 19.2 billion despite the COVID pandemic, making up 59% of the global ESCO market.[4]

As in case of energy audit, we do not have a global standardised format for ESPC. However, for Government sponsored ESPCs, country-specific standardised and model ESPC drafts are available. Generally, an seven-step process, as outlined below in Figure 5.8 is followed universally for ESPC projects.

The process is initiated by undertaking a walk-through audit of the host facility with a view to make a macro assessment of potential investment opportunity in energy savings measures. This service is provided by the ESCO at no cost to the host. ESCO also makes an assessment of the cost to be incurred for investment grade audit (IGA). An ESPC can be prepared only post completion of the IGA and agreement on the energy-saving measures to be implemented under the ESPC. ESCO, therefore, has to bear the development cost and risks till such time ESPC is signed. A 'Master Energy Savings Agreement (MESA)' is therefore signed post completion of the 'Walk-Through Audit (WTA)' to cover the development risk for the ESCO.

MESA remains in force during the development period till the time of signing of the ESPC. The roles, responsibilities, and obligations of the parties during the entire project cycle are articulated in the MESA. MESA articulates the obligation of the facility on compensation payable to the ESCO if the facility short closes or reduces the scope of the project prematurely. As an illustration, the facility is obligated to pay for the fee for energy audit in case the facility does not proceed further with implementation of the projects. Same will be the case if the facility decides not to implement some of the projects, which otherwise meet the technical and financial criteria laid down in the MESA.

As a corollary, ESCO will not be eligible for any compensation if it fails to identify energy-saving projects, meeting the technical and financial criteria, that would result in reduction of energy cost by say 10%.

ESCO remains singularly responsible for progressing the implementation of the projects post signing of the ESPC. This stage of pre-development of the projects offer opportunity to the project host to assess the skill competency and technical and financial capacity of the ESCO before signing of the ESPC. The relevant clauses from the MESA are later on incorporated in the ESPC and MESA is extinguished. The project is thereafter progressed by undertaking the follow-on steps as per the processes shown in Figure 5.8.

WTA: Walkthrough agreement, MESA: Master energy services agreement, IGA: investment grade energy audit, ESPC: Energy services performance contract, EPCM: Engineering, procurement & construction management, M&V: Measurement & verification

FIGURE 5.8   Eight-step process for ESPC.

OUTSOURCING MODEL

The outsourced energy management model is also sometime referred to as an energy performance management contract, or simply energy supply contract. This model represents a form of outsourcing in which the costs for all equipment upgrades, repairs, and so forth are borne by the ESCO, and the ESCO sells the energy output (such as steam, heating and cooling, or lighting) to the end user at an agreed price. Ownership of equipment may remain with the ESCO (Build-Own-Operate model) or may be transferred to the customer (Build-Own-Operate-Transfer). This model is being increasingly used in EU, China, and few more countries in Asia, such as India, Thailand for renewable and on-site power generation projects built by ESCOs. ESCOs provide electricity and/or heat as per mutually agreed contracted price of delivered energy linked to the fuel prices.

Development and construction of on-site "build-own-transfer" power-generating equipment using waste heat or byproduct gas from the manufacturing process located at the site is one of the most successful ESPC models operating in China in the metal sector. An ESCO constructs and operates the generating equipment, purchases the process energy resource (waste heat) for a small fee or no charge, and sells the electricity and or heat as steam from the WHR project to the facility owner usually at a rate below the cost of purchasing electricity from the utility grid.[5]

CHAUFFAGE/INTEGRATED ENERGY CONTRACTING

The model offers a tariff model for providing services to a facility as per agreed terms. Such services include space conditioning (degree days per square metre), maintaining the production process (TPD). The tariff model is mutually developed and incorporated in a supply and demand contract offered by the ESCO (Figure 5.9).

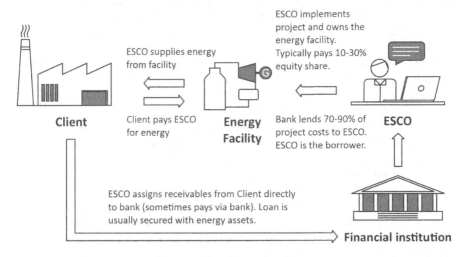

FIGURE 5.9   Chauffage model.

Potentially, there are two revenue streams for ESCO in this model, from sale of utility and from energy savings in the host facility from the contractual billing rate based on agreed unit price, dollar per square metre or dollar per tonne of production.

This concept derives from a previous contractual French approach of energy services delivered by a private company to a public authority or to another private body. In the former French approach, the contract used to consist of up to three elements designated as P1, P2, and P3, which are:

P1: Energy supply cost
P2: Maintenance cost
P3: Total guarantee cost (replacement cost of the equipment at the end of its life).

There are certain common threads that run across the three modes of implementation described above. The ESCO or the outsourcing agency carry out engineering, financing, procurement, construction, operation, and maintenance during the project tenure. The other common thread is higher level of energy performance by ESCOs, generally delivering over 25% energy savings (IEA).

However, there is one significant difference between the ESPC and the other two modes. The retrofits of the energy efficiency system become integral part of the existing system in case of ESPC. This introduces several complexities such as, engineering of the hook up, operation, and maintenance of the retrofits and the M&V system for capturing energy savings. In case of the other two modes, the new system generally remains at arm length from the existing system just as the electricity or gas utility suppliers to the facility. Further, it is not necessary to have a complex M&V system as performances delivered are clearly measurable and the facility does not have to bother about technical performance of the project.

The profile and structure of industry and the energy system therein are quite diverse. So are the country policies on implementation of energy efficiency in different sectors. ESPC by ESCOs is quite popular in the building segment in most of the countries. However, the development story is mixed for the industrial sector. As highlighted earlier, China has built the world's largest and fastest growing ESCO industry in the industrial sector (IEA). On the other hand, despite a healthy growth of ESCOs, the share of ESPC in the industrial sector is negligible in USA.[6] Outsource and Chauffage models are quite popular in Japan and Germany, the share being more than 50% in both the countries (IEA).

Even though ESPCs have become more standardised, there is still room for innovation to achieve superior energy performance. These innovations are often necessary to address specific concerns of a particular facility. The fundamental principles of risk sharing, financing, and M&V in ESPCs remain the same. However, the process and language of the ESPC contract can be simplified to bridge the knowledge gap between the ESCO and the facility engineers. This simplification helps facilitate a modular approach where a few projects are implemented initially, their success is tested, confidence is developed, and then the steps are taken for scaling up until all the projects discovered have been implemented.

Within the framework of the three key elements of ESPC (risk sharing, financing, and M&V), it is indeed possible to innovate the business process and financial mechanisms to suit a particular situation. Overall, ESPCs provide a framework for collaboration between ESCOs and facility owners to achieve energy efficiency goals. While the basic principles remain consistent, there is flexibility for innovation and customisation to enhance the performance and outcomes of ESPCs in different contexts.

The following case study illustrates one such example of innovation for implementation of an ESPC project in a paper mill in India.

Case Study 5.3: Innovative Business Model for Delivery of ESPC in a Paper Mill (Kuantum Papers Limited, India)

The energy system in this Agro-waste-based 80 TPD paper mill consisted of a cogeneration plant having one 50 TPH rice husk fired boiler and 1 MW back pressure TG set. The process steam demand was met from the cogeneration plant, whereas power requirement was sourced from the local utility. The plant was not able to operate the TG set due to operational and maintenance problem. The facility was aware of the potential for reduction of energy cost along with similar opportunities for improving productivity of the pulp and paper machines. The facility was somewhat under financial stress. They were looking for external support for financing of upgrade of the energy system. It was at that point of time that the author had launched ESPC business as an ESCO as part of a large Chemical manufacturing company. We offered to undertake the upgrade work under shared savings scheme without any upfront cost share by the facility. We carried out a walk-through audit and benchmark analysis based on which energy savings opportunities were identified as shown in Table 5.15.

TABLE 5.15
ESPC Target in a Paper Mill

Particulars	Units	'As is' Situation	'To be'	Savings Potential (%)
Boiler efficiency-specific steam generation to fuel consumption	T/T	3–3.2	4–4.5	35%
Revamp cogeneration system	MW	0	0.9	100%
Power distribution losses	kW	600–700	350–400	45%
Boiler auxiliary power	kWh/T	10	6	40%
Pumping system	kWh/m³	0.3–0.4	0.2–0.25	30%
Steam distribution	To be quantified post detailed audit			

The plant was not in a position to sign the MESA as it conflicted with terms and conditions from the lenders under which they were providing the working capital. We were advised to rank order the projects following two criteria:

Minimum capital need for implementation of the energy efficiency measures
Maximum benefits to ESCO under shared savings contract from a particular intervention that can compensate for the cost of detailed audit for the whole plant

Accordingly, the projects were rank ordered as shown in Table 5.16.

TABLE 5.16
Rank Ordering of Paper Mill Project

Project	Rank Order	Estimated Time for Implementation	Obligation of Paper Mill
Reduction of power distribution loss	1	2 days	Shut down of individual transformers for 4 hours each
Reducing boiler fuel consumption	2	7 days	Boiler efficiency test followed by shut down for 2 days and operational tuning thereafter
Reducing boiler auxiliary power	3	3 months	Detailed energy audit followed by engineering of projects
Reducing pumping power	4	6 months	Detailed energy audit followed by engineering of projects
Revamping cogeneration system	5	18 months	Study and upgrade of the feed water treatment system, condensate recovery, reinstallation of the boiler superheater, repair of turbine

It was decided to implement the first project as a pilot for enabling the facility and their banker to develop better appreciation of the ESPC process. The ESPC was negotiated amongst the facility, their banker and ESCO. Few innovative parts of the contracts are as follows:

Initial contract instead of MESA:

Facility will issue a normal work order to the ESCO for rectification of joints in one transformer LT chamber

Facility guarantees that ESCO will remain engaged for rectification work for the entire LT system should ESCO demonstrate success in the pilot. ESCO will be compensated should the facility short closes the contract

Bank provides letter of comfort to the ESCO on cash flow to ESCO

ESPC for the electrical distribution loss reduction

Entire work is service oriented except for consumables such as cable jointing materials

Savings sharing @ 50%, facility share to remain at that level, any shortfall to ESCO account

For higher savings over projected, share of facility @80%

Both parties will maintain an energy-saving account and 50% of the shared savings will be transferred to the same. The bank will have lien to this account as collateral for future borrowing for implementation of projects. This account will be extinguished on completion of the project tenure.

All the agreed projects will be implemented under self-financing drawing from the energy-saving account over a period of time so that cash flow on ESPC account remains positive, no additional cash is drawn from the operation and maintenance budget of the mill.

The pilot was implemented in 2 days' time. ESPC was then extended to cover all the transformers. Actual savings exceeded the guaranteed performance by over 50%. The facility was also able to derive co-benefit of higher production as the transformers could be loaded more.

Encouraged by the result, an innovative financing mechanism was developed for the execution of all the projects without straining the operational finance of the mill. A cash flow model was developed taking into account the capital expenses required over time for individual projects and the cash flow generated by the savings from previously implemented projects. Facility agreed to transfer their entire share of savings from previous projects to the capital account. Similarly, ESCO continued to transfer 50% of their share into the guarantee account. This way, all the projects were implemented in about 2 years' time with remarkable success:

- The paper mill achieved financial turnaround generating profits due to reduction in energy cost.
- The specific power consumption reduced from 1,200 to 1,300 kWh/T of paper to 950 kWh/T.

- The specific steam output from tice husk improved from 3.1 T/T average to 4.2 T/T.

In addition to the business success, the project taught us many new lessons that can be termed as co-benefits from the success of the business model innovated jointly by the three parties: the ESCO, the paper mill, and the bank.

For the ESCO:

Starting point of any ESPC should be far from the core process of a facility, removing the fear of failure

Training of ESCO staff on energy efficiency in a paper mill at no cost, extend the model to all forms of contract

Developing the confidence that ESCO can deliver projects better in a learning mode

For the facility:

Improving productivity of pulp and paper machines

Launching a long-term programme on energy and water efficiency

Significantly improving the environmental aspect of paper making by investing in resource recovery and pollution abatement technologies

Awareness on impact of energy efficiency on environment and monetising carbon savings

For the banks:

- Creating innovative instruments for financing of energy efficiency, water conservation and environment upgrade projects

The success of the project was also attributable to the leadership provided by the Director of the mill, Mr. Pavan Khaitan. This project was universally recognised as a successful demonstration of ESPC.

MEASUREMENT AND VERIFICATION

The importance of M&V in case of ESPC can never be over emphasised. Let us take the case of the first pilot project implemented in the paper mill as described above. How does one find out that the implemented project has reduced energy consumption? The project is selected based on temperature survey of the LT junction box of the transformer. How does one calculate the power loss due to high temperature? The power loss occurs due to voltage drop on account of resistance of the terminal joints and the current flow. Higher the resistance, higher will be the loss assuming current remaining constant. This resistance is dependent upon the quality of joint. Higher temperature occurs due to higher resistance on account of non-aligned and lose contact. The project is very simple, reducing the joint resistance, thereby reducing the voltage drop and hence power loss as per the following equations.

$$Q_1 = \sqrt{3} \times \partial V \times I \times \text{Cos } \Phi / 1{,}000$$

Q_1 is loss on account of voltage drop in kW; ∂V is voltage drop at the junction box in volt; I is running load in ampere; $\text{Cos}\Phi$ is the power factor.

Voltage drop is reduced by reducing the joint resistance by improving the cable terminations, improved contact area by proper alignment with the LT busbars and more contact pressure by proper tightening. The reduction of loss can be calculated by measuring resistance before and after improving the joint. However, it is practically not possible to measure the same as there are large number of cables that are connected to the busbars. Furthermore, the behaviour of the joint resistance changes with change in load on account of heating. We overcame the problem by measuring the voltage drop between the transformer LT terminal and LT panel terminals.

Next challenge is where and how to measure the voltage drops. Let us see how the transformer is connected and wherein we have access for field measurement (Figure 5.10).

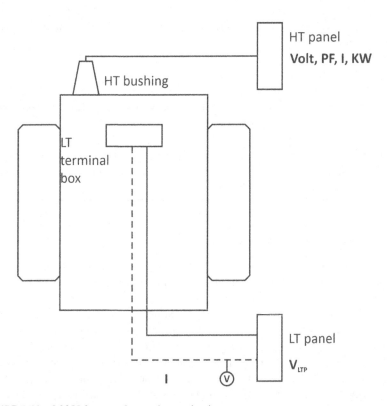

FIGURE 5.10 M&V for transformer loss reduction.

Voltage at the LT terminal box is function of transformation ratio and voltage drops in transformer measured at the LT busbar of the transformer. This is measurable by connecting a voltmeter at the busbar terminal. The metering system in both the HT and LT panels measures all the variables such as voltage, current, power, and power factor. Additionally, voltage is measured by connecting a voltmeter directly to the LT bus. Voltage difference between the voltage at the LT bus and LT panel account for voltage drop in the LT terminal box and the LT cable between the terminal box and the LT panel. The voltage drop in the cable is calculated by taking the resistance value of the cable and current flow

$$\partial V_{cable} = I^2 r$$

I is the current flow in the cable in ampere; r is the resistance calculated taking into account length of the cable and the vendor's data for cable.

$$\partial V_{ltsystem} = V_{lt} - V_{ltp}$$

$\partial V_{ltsystem}$ is the voltage drop between the LT terminal busbars and the terminal of LT cable.

Voltage drops in the LT terminal junction ∂V_{ltb}

$$\partial V_{ltb} = \partial V_{ltsystem} - \partial V_{cable}$$

Voltage drop is proportional to the load current, while power loss is proportional to the square of the current flow. Voltage drops as well as power loss needs to be determined at the baseline current. In case of change in the current flow during the M&V process, the actual voltage drops and power loss is to be adjusted for this change in the current flow.

The example above shows us the complexity of evaluation of retrofit energy projects. We therefore, need a well-designed and executed M&V plan for establishing the actual performance of a project. Considering the importance of the subject, a separate chapter (Chapter 7) has been included in this book with detailed deliberations on M&V protocols and plans for different types of projects.

CONCLUSION

In this chapter of the book, we have discussed the implementation issues that need consideration before undertaking execution of a retrofit project. The case studies have revealed how the configuration of a project developed during energy audit change as a result of more intensive technical and financial analyses.

Engineers are generally well conversant with the execution aspect of the project, engineering, procurement, and construction post approval. However, there is an intermediate step of project financing. Various steps for financial analysis have been explained as part of the case study. It is expected that the engineers and other professionals involved in developing energy efficiency retrofit projects will be able to use the tools and make better preparation of larger projects for financing of the same.

Various models for developing and execution of retrofit projects and the pros and cons have been explained. Different countries have adopted different models based on the historical experience with equal success. Every company can similarly adopt its own individual strategy so long performance goal is not compromised. However, for a bouquet of projects identified as a result of energy audit, ESPC offers a very attractive alternative. We have seen from a case study, how the ESPC model can be innovated to meet the specific need of a facility.

Despite the potential, ESCOs have not succeeded in fully tapping the market potential of ESPC in the industrial sector. This suggests that there are still many challenges or barriers preventing its widespread adoption. While ESPC may not have reached its full potential in the industrial sector, it has still provided significant indirect benefits. These 'rub-off' benefits have helped in improving sustainability of energy efficiency projects. Both industry and energy efficiency service providers have developed better understanding of the risks involved in execution of retrofit projects and approach for mitigation of the risks. Similarly, we have also developed deeper insights into the principles of M&V, which can be applied universally irrespective of the mode of implementation. This also suggests that M&V principles have become more universally applicable for evaluating project performance, even outside of energy efficiency projects.

The importance of M&V for energy efficiency retrofit projects can never be over emphasised. The case example of a project pilot has demonstrated the need for developing an M&V system even for a simple project like reduction of power loss in an electrical distribution system. This case example has been included to set the tone for the next chapter dedicated to M&V.

NOTES

1   Delphi A, Energy efficiency risk management, A report for financial intermediaries in Brazil, 2020, www.industrialenergyaccelerator.org.
2   IEA, Multiple benefits of energy efficiency, 2019, https://www.iea.org/reports/multiple-benefits-of-energy-efficiency/emissions-savings
3   USDOE, Superior energy performance, 2022, https://www.energy.gov/eere/iedo/superior-energy-performance.
4   IEA, Evolving energy service companies in China, 2021, https://www.iea.org/reports/evolving-energy-service-companies-in-china.
5   David Crossley, ESCOs in China, RAP, 2014, https://www.raponline.org/wp-content/uploads/2016/05/rap-crossley-escos-in-china-escowkshpbangkok-2014-nov-28-.pdf.
6   E. Stuart et al., US ESCO Industry-Industry size and recent market trends, (LBNL), 2021, https://www.osti.gov/biblio/1788023.

6 Measurement and Verification of Project Performance

INTRODUCTION

The measurement and verification (M&V) protocol plays a crucial role in establishing the actual performance of energy efficiency projects. It becomes even more important for retrofit projects that link payments to the service providers to the energy performance of the projects. It helps in accurately assessing and quantifying energy and financial savings as a result of implementation of the projects. Retrofit energy efficiency projects encounter multiple technical and financial risks. M&V protocol helps in identification of such risks and allocating the risks amongst stakeholders in a transparent manner.

By developing a well-defined M&V protocol at the inception stage of a project, potential controversies about actual energy savings post-implementation can be avoided, providing stakeholders with confidence in the project's success. A well-developed M&V protocol clearly defines the project boundary, establishes the baseline energy consumption, identify the variables impacting energy consumption, develop the adjustment principles, establishes the measurement plan and frequency for collection and validation of data and finally calculation of the energy savings. Clearly defining the project boundary is essential to identify the scope of the energy efficiency measures and the area or systems that will be impacted by the project. Establishing energy consumption baselines provides a reference point for comparison of energy performance before and after the implementation of the energy efficiency measures. Baselines can be historical energy data or assessed through mathematical models based on energy consumption patterns. It is also important to account for external factors that may affect energy consumption, such as changes in raw material quality, product mix or weather conditions. Adjustments and normalisation techniques help in isolating the impact of these independent variables and determination of actual savings due to implementation of the projects. The M&V protocol specifies the methods and equipment for measuring energy consumption accurately. Calculation routines are developed for determination of energy savings in a transparent manner based on reliable data.

Energy savings cannot be directly measured. It has to be calculated based on the measurement and analysis of data. The cost of M&V depends upon the desired level of accuracy in measurements as well as calculation rigours. The design of the M&V protocol for a particular project should therefore, be commensurate with the project capital investment and the energy and financial savings risks and allocation of risks between the service provider and the project host.

DOI: 10.1201/9781003415718-6

Overall, a well-designed M&V protocol provides a standardised and systematic approach to evaluating the success of energy efficiency projects. It enables stakeholders to have confidence in the reported energy and financial savings, ensuring the project's effectiveness and contributing to the broader goal of sustainable energy management. M&V protocols are generally followed for projects executed under Energy-Saving Performance Contracts (ESPC) or outsourced Engineering, Procurement and Construction (EPC) projects. However, M&V helps in better design and execution even for projects executed internally.

The key elements of a good M&V protocol consist of:

Evaluation and allocation of risks
M&V options and evaluation
M&V plan
Project-specific M&V
Measurements and analysis
Preparation of M&V reports

RISKS AND ALLOCATION

We have discussed about technical and financial risks encountered by retrofit energy efficiency measures in the previous chapter. We have also highlighted the risk score card and approach for derisking. Projects risks are well known, so are the arts and science of managing such risks. In fact, we have an ISO standard, ISO 31000 on risk management.[1] This standard provides principles, guidelines, and a framework of processes for managing risks. Such standards and similar other standards address issues faced in managing traditional projects. In such projects, risks are typically related to cost and time overruns, which can be mitigated through measures like liquidated damages and penalties. Similarly, performance risks are also covered through various guarantees and warranties, which remain valid for a limited period, usually a year from the date of commercial operation of the project. The cumulative impact of liquidated damages and performance penalty rarely exceeds 10% of the value of the project except in rare cases of catastrophic failures.

However, in retrofit ESPC projects under shared savings model, the magnitude of the risk is much higher as the entire investment made by an Energy Service Company (ESCO) remains at risk throughout the life of the project. The ESCO invests in the retrofit project, often by borrowing money from the lenders. Their revenue comes from the energy cost savings generated by the implemented projects. The cashflow from the saved energy needs to cover various expenses, such as operation and maintenance costs, taxes, repayment to lenders (if any financing was involved), and payments to the facility owner as per the shared savings formula before any cashflow reaches the ESCO.

In the context of M&V, risk refers to the uncertainty about the expected savings from the project both energy and financial and allocation of these risks amongst the project participants, for projects implemented under ESPC by ESCO.

The primary risk is failure of a project to deliver the projected energy savings. This can happen due to various factors starting with poor project design, choice of technology, poor construction, poor operation, and M&V error.[2]

Quality of design depends upon the quality of information obtained from the facility and validated during the energy audit. Information on process and equipment operating parameters collected during the audit must be error free. Sources of errors can be many such as field instruments, portable instruments, human interface (reading and entry), and plants operating points. Let us take the simple case of the water pump. The energy consumption in the pump depends upon the pump operating head, water flow, and the pump characteristics. How do we satisfy ourselves that the data collected during the audit represents the true operating condition and that the same will be there post implementation of the project. Data on operating parameters, pump head, water flow, and energy consumption over the measurement period will be obtained from the field instruments and/or portable instruments. Errors can creep in due to error in any of the instruments. There is also the larger question; do the data collected during this period truly represent the actual operating condition of the pump for the purpose of performance calculation?

Technology is usually selected in a consultative manner even though the responsibility rests with the ESCO. Similarly, we depend on the technology vendors for the performance parameters and calculation of energy savings. In case of the example cited above, the pump characteristics' curve will be obtained from different vendors. It is possible that a pump delivers highest efficiency at a pre-determined operating point while another pump is not as efficient at that point but delivers higher saving over the different operating points. M&V issue here is to establish that the calculated energy savings can be transparently related to the efficiency parameters as per the pump and system characteristics over the entire operating range. This helps in choosing the technology that may not be most efficient as per equipment name plate data but provides higher level of savings over the entire operating range. It is also possible that the investment cost is also lower for the second case, thereby improving the return on the investment.

Cost and time overrun does not normally impact the M&V plan directly. However, in case, the investment cost goes up, the financial savings may not be adequate to take care of the extra repayment obligation. Construction issue arises when the retrofit project requires mechanical hooking up with existing system, for example ducting for a waste heat recovery project or replacing a pump in a hydraulic network serviced by larger number of pumps. A poorly constructed duct will not only result in sub-optimal performance on energy savings but also can impact the productivity of the furnace. An M&V plan will clearly identify the risks involved in the retrofit hook up and articulate how the system parameters will be measured at the hook up points. Approach and methodology for measurements and analysis to take care of the uncertainties in calculation of savings will be clearly articulated. Operation and maintenance (O&M) can impact the project energy savings on two counts: loss of efficiency due to poor O&M and lower absolute value of energy savings due to lesser utilisation of the project.

The approach to the M&V activities is closely tied to the knowledge symmetry between the ESCO and the facility with such projects and the confidence level about the accuracy of the estimated energy savings. A less rigorous M&V approach will be acceptable for a project type with which the facility staff is familiar, such as replacement of a pump. It will not be so for a more complex project such as waste heat recovery wherein savings will be determined by calculation due to complexity of the measurement system In practice, a simple method for estimating payment risk might involve conducting an initial joint appraisal of the project's potential, evaluating the technical uncertainty and complexity of the project. Based on this assessment, projects can be categorised as per different risk levels, and M&V rigour can be tailored accordingly. Projects with higher payment risk would warrant more thorough M&V efforts to ensure accurate measurement and verification of savings. The allocation of responsibilities between the ESCO and the facility drives the M&V strategy for reducing the risks as well as fair allocation of the consequent financial impact on the parties. To allocate these risks appropriately, the responsibilities of the ESCO and the facility must be clearly defined. The facility often assumes responsibility for usage risk by agreeing to allow baseline adjustments based on measurements or accepting stipulated values for certain usage-related factors. Stipulations refer to the agreement between the ESCO and the facility to use predefined values for specific parameters throughout the contract term, irrespective of their actual behaviour. Using stipulations is a practical and cost-effective way to reduce M&V costs and allocate risks, especially when dealing with factors beyond the control of the ESCO. Properly crafted stipulations should not jeopardise the savings guarantee or the facility's ability to pay for the project. However, it is essential for the facility to understand the potential consequences of accepting stipulations, as they can shift some risk onto the facility.

Table 6.1 shows the summary of various risks from the perspective of M&V and allocation of these risks between the ESCO and the facility. Proper allocation keeps

TABLE 6.1
M&V for Risk Management

Risk	Primary Source of Risk	Allocation (Shared Savings) ESCO	Facility	M&V Intervention
Poor design of project	Poor quality information		√	Clearly developed and executed M&V plan for data collection
	Poor quality energy audit	√		
Wrong choice of technology	Inadequate analysis	√	√	Responsibility for technology selection, sign-off by stakeholders
	Wrong sizing	√		
Poor construction	Project boundary	√	√	Engineering drawings (P&ID) for hook ups-provisions of measurement points
	Poor engineering	√		
	Poor staffing	√		
Operation and maintenance	Poor staffing	√	√	Frequency of M&V, training, digitalisation, monitoring
	Inadequate training	√	√	
	Equipment degradation	√		
	Lower loading			
M&V plan and errors	Poor M&V plan	√	√	Acquisition, archival and retrieval of data, baseline adjustment, periodic calibration of instruments
	Data quality	√	√	
	Faulty meters	√		
	Modelling error	√		Periodic review of algorithms for modelling

the M&V cost at an acceptable level and at the same time maintaining confidence in the energy saving estimates.

The most reliable and accurate source of information is direct measurements. If available, data collected from short-term logging or existing data acquisition system (DAS) records can be used to develop models that estimate energy savings over time.

When direct measurements are not feasible or are costly to obtain, manufacturers' data or standard tables can be utilised. Manufacturers provide performance curves for equipment like pumps, fans, and many other energy conversion equipment. These curves depict the relationship between equipment performance and various operating parameters (e.g., flow rate, pressure, or temperature). By using these curves, one can estimate energy savings based on specific operating conditions.

In cases where manufacturer's data are not available, industry-accepted performance curves can be used. These curves are generally developed based on aggregated data from various sources and represent typical performance characteristics for certain types of equipment.

It's crucial to base these estimations on reliable and documentable sources to ensure a high degree of confidence in the reported savings over the entire post-installation period. Accurate estimations are essential for mitigating the risks, facilitating the decision-making, and ensuring success of energy efficiency projects.

M&V OPTIONS

An ideal situation from the M&V perspective will be where all the input and output parameters impacting energy consumption of an energy efficiency project can be directly measured as shown in Figure 6.1. The project involves replacement of an inefficient water pump.

Assuming hydraulic power requirement (function of pump delivery pressure and flow) remains the same for both baseline and project conditions, the difference in the energy consumption (calculated from the energy meter readings before and after) over a pre-determined period of time will represent the energy saved over that period. It is possible that the flow requirement and consequently the pressure and

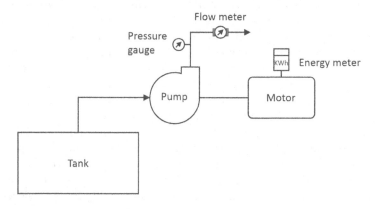

FIGURE 6.1   M&V of a water pump project.

hydraulic energy requirement vary over time. With current level of digitalisation, it will be possible to capture the impacts of variations, carryout adjustment for baseline and determine the saved energy the same way.

What happens if the project involves replacement of large number of pumps or replacement of pumps with modification of the piping and hydraulic network system. We will face major challenge, both technical and financial. Technical challenge will be to design a metering system that will capture the impact of changes in the hydraulic network on energy consumption. It will also be quite costly to provide metering system for all the pumps and the piping system. There could be situation where the cost of M&V system will be more than the cost of the energy-saving project. One has to therefore, find out compromised solution. Such solution evolves around partial measurement and developing engineering spreadsheets for calculation of power consumption and savings. Different scenarios were analysed with a view to develop measurement and calculation protocols for commonly acceptable approach for the estimation of actual energy savings.

An international protocol named 'International Performance Measurement and Verification Protocol (IPMVP)' has been developed with a view to meet the M&V challenges in various types of energy- and water-saving projects. This protocol has been developed with voluntary participation of energy efficiency professionals from across the gloves. Four different M&V options are the critical components of this protocol summarised in Table 6.2. The pros and cons of the options are what the author has faced while implementing ESPC projects in the large industries.

These options are for guidance only and not to be considered as standards. In practice, these options often require some sort of creative modification for increased transparency and hence, acceptability between the ESCO and the facility. One such creative modification is application of a concept called 'deemed savings'. This concept has been widely used in government and municipal energy efficiency projects for quite some time. Deemed savings refer to the predetermined saving values assigned to specific energy efficiency measures.[3] These values are developed based on commonly accepted data sources and analytical methods and are maintained by states or regional bodies.

The purpose of using deemed savings is to simplify the evaluation of energy efficiency projects and to streamline the process of calculating energy savings. Instead of conducting costly and time-consuming individual measurements and verification for each project, deemed savings allow project developers and policymakers to rely on established values for energy savings associated with particular efficiency measures.

Taking cue from this commonly used definition for deemed savings and the IPMVP M&V Option A, it is possible to significantly eliminate the uncertainties about actual savings at a low cost. We can illustrate this with the specific example of water pumping project cited above. Post completion of the efficiency upgrade project, we can determine the actual efficiency of few pumps at two to three different operating points. We can carry out the efficiency tests for a longer duration to capture various operating points. We can then calculate the annual energy saving based on the efficiency tests and historical data on pump loading. Overall energy savings (deemed savings) for the entire system is then calculated based on projected efficiencies and pump loading.

TABLE 6.2
IPMVP M&V Options

Options	Description	Merits	Demerits	Typical Application
A-Partial retrofit isolation	Savings are estimated based on partial measurements and assumptions for certain parameters	Very easy and low cost of M&V	Lower level of acceptability particularly when responsibility for operation control is not clear and doubt about the replicability based on sample measurements	In the case of the pump illustrated above, we can use this option by extrapolating overall savings from larger number of pumps by measuring the savings from few sample cases using portable instruments
B-Retrofit isolation	Same as above but wider coverage of the entire system impacted by the project	Robust and accurate	High cost of metering and monitoring	High investment projects, where high cost of metering can be justified
C-whole facility	Energy savings can be directly determined by actual measurements of inputs and outputs	Most accurate. Results and impact can be transparently established. Easy to carry out baseline adjustment	Difficult to implement in retrofit applications particularly where inputs and outputs cannot be specifically linked to the ESPC projects	Waste heat recovery, cogeneration. Utility outsourcing. Utility systems like pumps, compressors, lighting, etc.
D-Calibrated simulation	The energy savings are determined based on pilot study and applying simulation methodology for application to the whole facility or sub-facility	Reasonably accurate system can be developed for determination of energy savings by periodic test and performance analysis	Requires higher skill for carrying out simulation. Information asymmetry can create problem of acceptability	Mainly in buildings

However, to ensure accuracy and credibility, it's essential to carry out these efficiency tests and calculations using reliable methodologies and standards, such as IPMVP Option A. This option involves monitoring and verifying the performance of the individual pumps or the overall pumping system using calibrated meters and instrumentation. Comparing the savings calculated based on deemed savings principles against the savings calculated from installed meters helps to reconcile and validate the accuracy of the deemed savings approach.

By doing so, we can significantly reduce uncertainties about the actual savings achieved through the efficiency upgrade project at a relatively low cost compared to more complex and expensive M&V approaches.

It's important to note that while deemed savings can be a valuable tool, it is not suitable for all projects and situations. Their effectiveness depends on the availability of historical data, accurate engineering calculations, and the suitability of the deemed savings model for the specific project being evaluated. We would further analyse the pros and cons of applying deemed savings concept for different kinds of retrofit projects while deliberating on Project-specific M&V.

M&V PLAN

The M&V plan is one of the most important documents that is developed prior to undertaking the implementation of energy-saving projects. This plan establishes the basis for determination of actual energy savings from individual projects and overall. A good M&V plan ensures transparency of the processes and the quality and credibility of achieved outcomes. It helps in reducing the risks of project failures, establishes the roles and responsibilities of the stakeholders for long-term sustainability of the energy savings achieved post project implementation. This plan also sets the framework for preparation of the M&V report post commissioning of the energy-saving projects. The extent and rigours of the M&V activities are planned considering the project complexity and expected level of savings. The amount of detailing required for a M&V plan depends on the scale of the project and costs associated with it. A less-accurate and less-expensive M&V plan is better than an incomplete or poorly executed M&V plan, which is more complex and theoretically more accurate.

An M&V plan at the minimum should include[4]:

List of energy savings projects
Project boundaries
Baseline energy consumption, adjustment of baseline
M&V options
Instrumentation, measurement, and data management plan
Preparation of M&V reports

LIST OF PROJECTS

The energy audit report recommends a list of projects for implementation. Following this, the facility engineers prepare more detailed feasibility report

TABLE 6.3
Project Types

Serial Number	Energy Savings Project Types
1	Replacement of water pumps
2	Installation of VFD and Automation for high-capacity pumps
3	Replacement of Steam Driven CO Gas Exhausters with electrical drive

based on which the projects are screened for implementation utilising different modes of implementation as has been explained in Chapter 5. The M&V requirement for these projects are assessed depending upon the magnitude of savings and investments and measurement complexities. Let us take the case of a project which has three different project types as shown in Table 6.3 and develop an M&V plan around them.

Project Boundaries and Performance Period

The project boundaries for most of the listed projects above will be the electrical terminals at the power control or motor control centres serving the particular motor and the delivery nozzle of the pump as illustrated in Figure 6.1. However, for the project 3, we will have two sets of boundaries, one for the determination of baseline energy and the second for the project. The baseline boundary for energy input calculation will comprise of the steam inlet point at the inlet of meter or the turbine nozzle and the steam exhaust nozzle of the turbine. The boundary for the project will be the inlet terminal of the motor that has replaced the turbine. The project output will be exhauster flow, which will remain the same for both baseline and project. In many cases, it is possible to have larger project boundary consisting of other equipment connected to the system, for example, number of other exhausters performing similar functions. The input energy measurement points under such case may shift to common steam header and a common PCC or MCC for the motors. This logic can be further extended to cover all the equipment serving a particular section, say rolling mills, where section-wise specific energy consumption is monitored as per set performance indices. Furthermore, the industries are increasingly required to reduce the specific energy consumption for products in line with targets set under the 'net zero' scenario. This will mean the M&V protocol has to be designed to demonstrate that the overall specific energy consumption (SEC) has been reduced. Therefore, even if the project pertains to a section, result from the project has to show the impact on overall SEC. Figure 6.2 shows the impact of the savings project on reduction of overall SEC.

This means the project boundary has been shifted to the overall energy consumption, though actual project island is only a few sections of the plant. This has major impact on determination of baseline energy consumption and preparation of M&V report linking project savings to reduction of SEC overall.

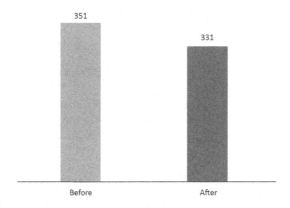

FIGURE 6.2   Reduction of SEC.

The project performance period is mutually decided based on the payback period. M&V activities over the performance period includes the periodicity of measurements and verification and content of the periodic M&V reports.

BASELINE ENERGY CONSUMPTION, ADJUSTMENT OF BASELINE

Baseline energy consumption refers to the amount of energy that is typically used by an equipment or a system, or a process or a facility under prevailing operating conditions prior to implementation of an energy-saving project. It serves as the reference point for measuring energy efficiency improvements. Baseline energy consumption can be related to two sets of variables, dependent and independent. Dependent variables (also called controllable factors) are the ones that we consider as factors that we can control for improving the system efficiency within the project boundary. In case of the pumping system, these may include factors such as pump efficiency, throttling losses, and automation. Then we have the independent variables (uncontrollable factors) that are not within our control but have impact on the energy consumptions. These may include weather, operating hours of the pump, changes in process demand, etc. In the context of an industrial facility, it is very important to identify the independent variables and accurately determine their impact on SEC. In a production plant, for example, the quality of raw materials, production capacity utilisation, product mix are independent variables that impact SEC. It is therefore, necessary to study the production process and energy consumption in details with a view to assess the impact of all independent variables. Baseline energy consumption has to be adjusted factoring the impact of these independent variables. The goal of adjusting the baseline is to create a fair comparison between the current energy consumption and the reference point established by the baseline. This helps accurately assess the effectiveness of energy-saving measures and provides insight into whether changes are having the desired impact. For this, it is necessary to identify the independent variables that can impact the savings. Table 6.4 shows a list of typical independent variables in the energy intensive industries.

TABLE 6.4

Independent Variables Industries

Sector	Independent Variables
Aluminium	Capacity utilisation, aAnode sourcing, current density, process technology
Cement	Limestone quality, fuel mix and quality, process technology, capacity utilisation, product mix
Chlor-alkali	Capacity utilisation, cell current density, cell technology, membrane/electrode life
Nitrogen fertiliser	Process technology, fuel/feedstock, plant vintage, capacity utilisation
Iron and Steel	Process technology, product mix, coal quality, other charge materials quality, capacity utilisation, source of power and heat, and energy accounting system
Pulp and Paper	Pulping technology, raw material usage, capacity utilisation, product mix, source of power and steam

Capacity utilisation (CU) is the most important factor amongst all the independent variables for all the industries. Baseline adjustment factors are determined for every individual plant by carrying out statistical analysis based on historical data. Actual savings from projects are determined based on overall savings net of baseline adjustment as shown in Figure 6.3.

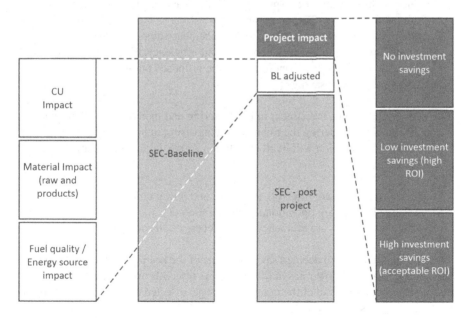

FIGURE 6.3   Baseline and adjustment.

It's important to use consistent and well-documented methods for baseline adjustment to ensure the validity of energy efficiency evaluations. Energy audits, historical data analysis, and mathematical modelling are common approaches used to establish and adjust baseline energy consumption.

M&V Options

Various M&V options, their applicability and the pros and cons have been discussed. We have to choose the M&V option for every individual project after carefully analysing the cost against the financial gain due to energy savings. Let us take the case of the project 1 on replacement of spray pump listed in Table 6.3. We can choose from any of the three options A, B, and C for this. Technically, C is the best option. It is possible to incorporate a digital system that will compute the energy savings as well as SEC for the project. It can also be configured for establishing co-relationship of energy savings from the project to the overall SEC for the facility. This satisfies the important considerations of transparency, accuracy, and verifiability. The metering and monitoring system will be required for

> In situ measurement of delivery pressure, flow and input electrical energy at several operating points that reflect actual operating conditions (for over 80% of the time as informed by plant)
> Calculation of hydraulic power and determination of SEC and efficiencies at those operating points
> Developing histogram for computation of annual average flow and energy consumption (Baseline SEC and efficiency)
> Obtain pump characteristics curve for replacement pumps
> Computation of annual energy consumption, efficiency, SEC based on histogram data on operating conditions for the proposed pump
> Estimation of energy savings

This will require major investment in the metering and monitoring system that may not be justifiable considering the financial savings from improved efficiency.

A compromise solution will be to use Option A, partial retrofit isolation.

In this case,

> Measured values are flow (Q), head (H), and power consumption (P)
> Assumed values are operational time duration (T)
> Calculated values are efficiency (η) and energy consumption

This option is also called deemed savings as one of the key parameters that is overall time duration as well flow rates at different time intervals are assumed. This option also assumes that the head of the pump will not get affected due to inadequate maintenance of the system.

The option will not require any additional instrumentation as portable instruments will be used during audit as well post implementation of project for measurement of flow, head, and power. Baseline SEC will be calculated based on measurements during audit. Similarly, project SEC will be calculated based on measurements post implementation of projects. Energy savings will be calculated assuming the flow as per historical data.

For some project types, it will be difficult to exercise options A, B, and C due to various constraints such as complex project boundary, measurability of key parameters, and computation of energy savings. In this context, let us examine the project 4 from Table 6.3 on list of projects. The project is replacing steam drive by electric drive. Steam drive consists of back pressure turbine the exhaust steam being used for process heating. We will have to find another source for process steam that will not adversely impact the system heat balance. So, the project boundary will extend to the process utilising the exhaust steam. We will have to establish the baseline energy consumption by measuring the steam pressure and temperature for both inlet and exhaust steam and the steam flow capturing the variabilities. The next issue will be determination of the conversion factor for steam enthalpy to power. We will calculate the enthalpy drop in the turbine for calculating the baseline energy consumption and then use the conversion factor 860 kcal/kWh for expressing baseline energy consumption in kWh. However, the heat rate for power generation in the system depends on the mix of technologies – renewables such as hydro, solar, and wind and non-renewables such as thermal. We can take the factor at 860 kcal/kWh for the power sourced from renewable system. But for the rest, which is sourced from thermal system, the conversion factor will depend upon the aggregate of heat rate for different thermal power units. The figure can range from about 1,500 kcal/kWh for high-efficiency combined cycle to about 2,450 kcal/kWh for an efficient coal fired thermal plant. The ratio of renewable to thermal will keep on increasing for obvious reasons. Similarly, aggregated heat rate of thermal plants is likely to decrease with decommissioning of inefficient plants. It will therefore, not be prudent to assume a fixed heat rate for calculation of lifetime savings. In such type of project therefore, we will be better off with Option D, simulation and engineering calculation for estimation of energy savings. The baseline document needs to therefore, must highlight not only the rationale for choosing the option but also include the assumptions used for savings calculation.

Instrumentation, Measurement, and Data Management Plan

The plan for instrumentation and measurement is correlated to the selected M&V option. As stated above, in case of deemed savings option, it is possible to avoid on-line instrumentation. Energy savings are calculated based on periodic measurement of the relevant parameters using portable instruments. In case, any of the other three options (Option A, Option B, and Option C) are chosen, on-line metering, logging, measurement, and computational systems will be required. The extent of requirement will be least for Option A while it will be highest for Option C. Table 6.5 illustrates the measurements requirement for the three options. Metering and measurement will capture the data on pump head and flow and input power while computation system will be required for calculation of SEC, energy savings, and baseline adjustment.

TABLE 6.5

Metering and Computing Plan-Replacement of Pump

Option	Metering and Computing for M&V				
	Indicator	Logger	Integrator	Computer	Reporter
A-Partial retrofit isolation	Yes	No	No	No	No
B-Retrofit isolation	Yes	No	Yes (Power)	No	No
C-Whole facility	Yes	Yes	Yes	Yes	Yes

The plan will also include the system for archival and retrieval of data, medium of storage (electronic/paper), frequency for archival, period of storage (which data for how long), and responsibility for archival over the project performance period. In case of engagement of third party for verification, witnessing requirement should be clearly spelled out in the plan.

M&V REPORTING REQUIREMENT

The level of detail and complexity of reporting can vary significantly based on the implementation mode and the specific programme's goals.

In government-supported energy efficiency performance contract projects, especially in the building sector, the reporting requirements are extensive and cover all stages of the project, from the initial energy audit to the verification of energy savings. The focus here are transparency, accuracy, and accountability due to the involvement of public finance. Detailed guidelines on M&V reporting requirements for such programmatic projects are available in the respective public domains.

For similar programmes in the industrial sector, in few countries like China and India, there are mandates that outline the specific reporting requirements for these projects. These guidelines provide instructions on how to prepare M&V reports that adhere to the programme's standards and goals.

On the other hand, under voluntary programmes, the reporting requirements are generally less elaborate to keep the cost at reasonable level. It is more focused on demonstrating the achieved energy savings and comparing them against projections. For voluntary projects therefore, the reporting requirement should be spelled out as part of the M&V plan. For the others, just a clause confirming that the reports will be prepared as per mandated formats should be adequate.

M&V OF PROJECTS

M&V of every individual project is carried out as per the M&V plan for the particular project. Let us demonstrate this for the project 1 on replacement of spray pump. Energy is saved by replacing the identified inefficient pump with a more efficient one, the dotted line showing the project boundary (Figure 6.4).

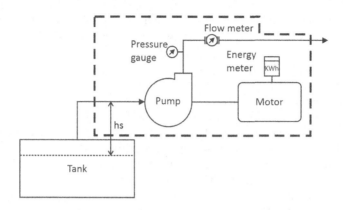

FIGURE 6.4 Project boundary for M&V.

The boundary limits are the suction and delivery points of the pump and electrical connections at the motor terminals. We will use the general equation for pump:

$$P_h = Q_{av} \times (h_d - h_s) \times \rho \times g/(3.6 \times 10^6)$$

where P_h is the hydraulic power generated by the pump in kW; Q_{av} is the flow in m³/h; h_d is the pump discharge head in M; h_s is the suction head in M; ρ is the density of the fluid (996 kg/m³ for water at ambient temperature); g is the gravitational acceleration (9.81 m/s²).

We will calculate the pump efficiency and energy consumption for both baseline and project by measuring the required parameters and using the pump equation as per the following protocol.

Performance testing: Once post commissioning and

Once before handing over after one year's operation

Baseline energy consumption: Measured for four hours duration

Measured values-suction and discharge heads, flow and electrical energy
Calculated pump efficiency, SEC, energy savings

Assumptions: Operating hours for calculation of annual energy savings (O_h/year)

Baseline adjustment factor: Change in pressure head due to operating/maintenance conditions other than flow related

M&V option: Option A, partial retrofit isolation

Instrumentation, measurement and data plan: Portable instruments-digital pressure recorder, ultrasonic flowmeter, power analyser and stopwatch (all instruments with calibration certificate) with datalogging and computer interconnection facility. Data captured electronically will be uploaded for open sharing with designated engineers.

Energy-saving calculation: Based on measured and calculated data, similar tests will be performed for the determination of the baseline parameters as well as during M&V post commissioning and follow on tests as per agreed frequency. Measured values, assumptions, and the calculation methods are to be agreed and protocols included for every project, illustrated for the pump project as follows.

Duration of test − T Hours
Average water flow − $Q_{av} = Q_t/T$ in m³/h
Average head-$h = h_d − h_s$ in M
Hydraulic power $P_h = Q_{av} \times h \times 996 \times 9.81/ (3.6 \times 10^6)$ in kW
where Q_t is the measured flow of water during test duration, h is the pump net head in M & P_h is the calculated hydraulic power in kW.
Measured energy consumption at motor terminals during test = E_t in kWH
Calculated SEC = E_t/Q_t in kWh/m³
Calculated baseline $SEC_{bl} = E_{tbl}/Q_{tbl}$ in kWh/m³
Calculated project $SEC_p = E_{tp}/Q_{tp}$ in kWh/m³
Reduction in $SEC_{sav} = SEC_{bl} − SEC_p$
Annual energy savings = $SEC_{sav} \times Q_{av} \times O_h$/year in kWh/year
Calculated input power $P_i = E_t/T$ in kW
Pump unit efficiency $\acute{\eta} = P_h/P_i$
Motor efficiency = 0.9 (Nameplate)
Pump efficiency $\acute{\eta}_p = \acute{\eta}/0.9$
Pump efficiency figures would be computed for baseline and project to determine efficiency gain

Energy savings have been determined assuming no change in the flow rate (Q_{av}), from baseline condition. In case of change in the flow rate, calculated energy savings will also change less than the projected savings. This will impact the allocation of shared savings for project implemented under performance contract. Under the deemed savings concept therefore, the flow rate measured during project M&V is used as an input data for calculation of project SEC only and not actual savings. Actual savings is calculated taking into account the baseline flow rate Q_{av} irrespective of actual flow rate during project M&V.

Actual savings figure may also require adjustment due to change in the pressure head on account of factors beyond the project boundary, such as scaling in the pipeline. In case of such change, the calculated savings figure will be adjusted by a factor. This factor is the ratio of the baseline pressure head to the project pressure head.

In summary, this text outlines the methodology for calculating energy savings and efficiency improvements related to the pump performance only, particularly in the context of an ESPC.

Basic principles for calculation and determination of energy savings for different project types using M&V options A, B, and C remain the same. However, we use less assumptions and more of data for Option B and even more for Option C. Instrumentation requirement therefore, increases. M&V Option D is mostly used in the building sector as it is difficult to measure many of the parameters influencing energy consumptions.

M&V REPORT

The focus of M&V report is presentation of the calculated and verified energy savings achieved in a summary form followed by reports on various actions taken as per the M&V plan. In case of deviations from the plan, the reasons for the same and plan for future are detailed out. The report also provides summary recommendations on corrective actions in case of shortfall in energy savings from the projected values. As stated earlier, the reporting requirement for government-supported and mandated projects are standardised.

The structure of the M&V report for bilateral projects can be a simple summary table on energy savings or an extended version as per the M&V plan. The extended version of an M&V report usually consists of:

An executive summary of findings from the post project M&V activities
The approach and methodology for the M&V
Summary table-projects
Data, measurement, monitoring
Recommendations

EXECUTIVE SUMMARY

The executive summary is prepared capturing the project highlights for quick appraisal of the project performance. The contents of the executive summary include a brief narration of the implemented projects, approach and methodology for M&V, a table on project performance and summary of recommendations for improvement of project performance as well as M&V process.

APPROACH AND METHODOLOGY

The specifics of the approach and methodology vary depending on the industry, the spirit behind the M&V plan, the complexity of the processes, and the size and complexity of the projects. It's essential to tailor the M&V approach to the unique characteristics of each industrial energy efficiency project. A well-structured M&V approach enhances credibility, helps identify areas for improvement, and supports decision-making for future projects. This is achieved by undertaking the various tasks in a systematic way and reporting on specifics such as:

Period for the M&V, dates, participants
Field and office activities-data collection, datalogging, analysis, presentation
Instrumentation and measurement systems-on-line and portable
Calculation on savings and adjustments
Results and interpretation
Stakeholder' consultation

Stakeholders' consultation is a critical requirement for the collective analysis of the project performance and fixing responsibilities for corrective actions, if required. Minutes of the stakeholders' consultation meeting forms an integral part of the M&V report.

SUMMARY TABLE – PROJECT PERFORMANCE

The actual performance for every individual project is tabulated as illustrated for a particular project in Table 6.6.

TABLE 6.6
Project-wise M&V Report

		Energy Performance-kWh/Year				Deviation		
Projects	M&V Option	Baseline	At M&V	Baseline Adjustment	Adjusted savings	Projected Savings[a] kWh/year	Deviation kWh/year	% Deviation
VFD for fan	D	825,000	562,500	104%	585,000	600,000	15,000	2.5%

[a] As per energy audit or project report

Option D has been used as engineering calculation was used to determine the savings based on several spot efficiency tests at different operating points and projected aggregate loading over the year. Baseline energy consumption was adjusted taking into account the total hours of operation in the year which was 7,800 h against baseline of 7,500 h.

In case of adverse deviation beyond a pre-determined agreed limit, say 2.5%, the reasons for the same has to be explained, responsibilities for corrective action highlighted as per minutes of the stakeholders' consultation meeting and savings share recomputed.

DATA, MEASUREMENT, MONITORING

A review report on adequacy and quality of data gathered against provisions in the M&V plan is highlighted. This evaluation helps in identifying instances of missing or incomplete data, impacting the accuracy of energy-saving calculations. Improved validation checks and data cleansing processes are carried out to enhance accuracy. Scope for digitalisation for enhancing the quality of data collection and monitoring are investigated and reported for sustaining energy savings achievements. Implementation of automated data collection systems that directly interface with energy-consuming systems are explored for reducing manual intervention and the likelihood of human errors. It is often possible to advance the algorithms to validate collected data in real-time. It also enables the development of predictive analytics to forecast energy consumption patterns based on historical data and external factors. It is possible to integrate emission factor databases with the monitoring system to calculate the emission reductions resulting from energy savings accurately. Feasibility of low-cost option for implementing data visualisation dashboards that present energy savings and emission reduction metrics in a visually understandable format are analysed. This facilitates communication of achievements to stakeholders.

Review and redesign of the monitoring system plays a critical role for the advancement of the energy efficiency journey for any organisation. A good quality report on the monitoring system lays the foundation for digitalisation and visualisation of achievement from individual projects and the facility overall. Visualisation of achievements impacts in a major way in increasing employee engagement for promoting energy savings.

RECOMMENDATIONS

The M&V process involves analysis of the risks and various other issues faced during different phases of the project, preparation of M&V plan, construction, commissioning and periodic M&V of performance. Discrepancies between projected savings and actual outcomes is the most critical issue that has to be analysed in micro details. Recommendations are then made to improve both project performance and the M&V activities themselves.

Deviations between projected and actual savings can occur due to multiple factors such as lower equipment performance, deficiency in O&M, measurement errors and changes in the operating conditions. It's important to assess the contribution of each factor to the deviations. This helps in understanding the root causes of the discrepancies.

An action plan is developed based on the analysis, and responsibilities are allocated to relevant parties for implementing corrective measures.

Sometimes, changes occur within the facility during project implementation that affect energy savings, the guarantee, and the M&V plan. The recommended measures in such cases include review of the project including the baseline, reassessment of project savings and updating of the M&V plan.

CONCLUSION

This chapter has highlighted the important role of M&V for transparently establishing the energy performance from investment in energy efficiency projects. Energy efficiency projects often underperform thereby risking the investment. The risks involved and the need for allocation amongst different stakeholders have been deliberated. The role of M&V for risk management has been explained. The process begins with preparation of an M&V plan post selection of the projects for implementation. The content of a typical industrial M&V plan has been provided along with approach for preparation of the plan. Different M&V options have been explained along with rationale for choosing a particular option. The need for data and the process of collection of the data under different M&V options have also been discussed.

The process of determination of the baseline and energy savings have been explained with an illustrative example. Through rigorous data collection, analysis, and comparison against baseline data, the tangible benefits of the energy efficiency initiatives are well understood by all stakeholders. The results demonstrate not only energy savings but also reduced operational costs and environmental benefits, all of which contribute to the long-term sustainability of the organisation.

The M&V process has also highlighted the importance of continuous monitoring and improvement. By regularly tracking energy consumption patterns and performance metrics, the organisation can identify deviations, address potential issues, and fine-tune strategies for even greater efficiency gains. Furthermore, the detailed findings presented in an M&V report serve as a valuable resource for future decision-making, guiding investment in similar projects and ensuring maximum return on investment. Integrating emission factor databases with the monitoring system has been suggested for calculation of the emission reductions resulting from energy savings accurately.

It is important to recognise that industrial energy efficiency is an ongoing journey, and the success achieved through a project sets a precedent for the commitment of the organisation towards a greener and more sustainable future.

Digitalisation and visualisation play a very important role in monitoring project performance. Benefits of integrating digitalisation in the M&V plan and report have been highlighted.

The content of an M&V report along with approach and methodology for preparing the report vary depending upon the mode of implementation. When projects are implemented due to regulation or mandates, the M&V report is prepared as per set protocol. However, for industrial projects, which are mostly bilateral in nature, the M&V report is prepared as per jointly agreed outline. The contents of the report for a typical such project have been detailed out.

In conclusion, M&V plan and M&V reports are the two most important documents required for monitoring of project performance. Success of the M&V process depends upon the quality of the M&V plan. Various aspects of the planning process have been detailed out in this chapter. The frameworks for both the plan and M&V report have been provided. These frameworks are meant to act as guide for the readers to help them prepare project-specific plan and M&V activities.

NOTES

1 ISO, ISO 31000 Risk management, 2018, https://www.iso.org/iso-31000-risk-management.html.
2 W. L. Lee et al., Risks in energy performance contracting (EPC) projects, 2015, http://getrede.ca/wp-content/uploads/2017/05/Lee-2015.pdf.
3 M&V Focus, EVO, 2019, https://evo-world.org/en/news-media/m-v-focus/868-m-v-focus-issue-5/1154-evo-deemed-savings-statement.
4 Measurement and verification (M&V) guidelines for the energy efficiency fund (E2F), 2019, https://www.nea.gov.sg/docs/default-source/default-document-library/guidelines-on-mv-plan-and-mv-report.pdf.

7 Employee Engagement for Sustainability

INTRODUCTION

Professionals involved in any business performance improvement programme play multiple roles as a doer, as a manager, and finally as a leader depending upon her position in the organisation and the requirement of the project. Many of us would recollect the roles played by us as doer during the initial phase of participation in an industrial energy efficiency programme. I still recollect my first assignment of implementation of a return condensate energy efficiency project as a young trainee engineer. The job consisted of installation of condensate collection bottle in one of the chemical processing sections and pumping the collected condensate to the deaerator tank of the boiler. The impact of the project on overall energy consumption was insignificant. I was chosen from amongst the young engineer trainees for the assignment; the importance of the job, the way to execute it, approach for measurement and calculation of the efficiency gain was explained to me in a meeting attended by senior management team. Finally, the result from the project helped me obtain high performance score resulting in early confirmation. I learned about arts and science of employee engagement much later but for me this still remains as one of the classical examples of empowering employee through a process of engagement.

Employee engagement is a process of inclusion of the relevant employee in a programme depending upon the expected role to be performed by her in the project. It can be just keeping her informed on the development to participation in the decision-making process. A formal engagement process helps in resolution of conflicts by addressing the concerns and interests of the employees involved during the different phases of the project cycle. This process deals with the behavioural aspect of the individual employees. The arts and sciences of engagements are learned from the experiences gained by individuals dealing with different types of projects, organisations and individuals. In the context of energy efficiency, aligning the organisation is a multi-step process. And different organisations have adopted different models with equal success.

Key Stakeholders, Their Roles, and Concerns

A stakeholder is defined as anybody who can affect or potentially be affected by any new initiative such as energy audit in a facility. All individuals engaged in the operation and maintenance of a facility are key stakeholders for conducting energy audit and implementation of efficiency improvement measures. Their actions influence the outcome of the audit and at the same time, findings from the audit can affect their interests. Energy and cost savings, though important, may not be considered

DOI: 10.1201/9781003415718-7

important enough for attracting attention of the senior leadership except in rare cases of mandate or competitive pressure. We will be essentially dealing with middle management during the audit. The middle management by nature feels most insecure in any industrial organisations. Seldom ever, they get credit for success of new initiatives but have to bear the brunt of responsibility in case of failures. This makes them feel vulnerable to any changes. They take defensive positions blocking new initiative such as energy audit. For example, a facility manager may raise the question of reliability of operation as the key concern, whereas in reality, it could be his perception about threat to his job in case the auditor is able to show the way for reducing the energy consumption. Furthermore, he may also feel that his work load will increase as a result of changes. On the other hand, he does not see any tangible benefit for him from the reduction of energy cost. The process of knowing his key concerns and allaying his threat perception would be one of the key objectives of the engagement programme. Inability to identify any key stakeholder in the beginning and not analysing his concern can create major hurdle in closing a project at a later stage. Similarly, there would be other stakeholders who may not be directly impacted by the finding of the energy audit but have concerns about the impact on their performance.

Categorising stakeholders based on their role and influence is a common practice in project management and decision-making processes. It helps in understanding their potential impact and involvement in the project. Classifying stakeholders into five broad categories – decision maker, influencers, implementers, supporters, and opponents – provides a useful framework for understanding and managing their roles and interests in a project.

> **Decision Makers**: These stakeholders have the authority to make final decisions regarding the project. They hold the power to approve or reject proposals, allocate resources, and determine the overall direction of the project. Decision makers can include senior executives, board members, or high-level regulatory officials.
>
> **Influencers**: Influencers may not have the final decision-making authority, but they hold significant influence and can shape the project's outcome. They possess expertise, knowledge, or a strong network that can sway decisions. Influencers can be subject matter experts, consultants, industry leaders, or key opinion leaders.
>
> **Implementers**: Implementers are responsible for executing and implementing the project activities. They are directly involved in the day-to-day operations and play a crucial role in ensuring project success. Implementers can include project managers, team leaders, department heads, and frontline workers.
>
> **Supporters**: Supporters are stakeholders who may not directly participate in decision-making or implementation but provide resources, assistance, or advocacy for the project. They can be internal or external stakeholders who believe in the project's objectives and contribute to its success. Supporters may include employees, unions, community groups, or non-governmental organisations.

Opponents: Opponents are stakeholders who have concerns, objections, or opposing views regarding the project. They may resist or challenge the project's implementation due to various reasons such as conflicting interests, potential negative impacts, or alternative priorities. Opponents can include internal and competitors, regulatory bodies, environmental groups, or community members with dissenting opinions.

Consulting organisations engaged in energy audit or implementation of energy efficiency projects need to study the host organisation dynamics to identify and classify the stakeholders early in the project. This helps them in designing the communication strategy for smooth conduct of the assignment. The entire process has to be conducted in a discrete manner considering the sensitivities of every individual stakeholder.

ROLE OF LEADERSHIP

It is a well-recognised fact that the success in implementation of energy efficiency measures depends largely upon the involvement of the leadership. Until recent past, it was rare to see active involvement of senior management in activities relating to energy efficiency. It used to be a common refrain amongst energy professionals on how to push up the energy efficiency agenda from the 'boiler house to the board room' of 'shop floor to the top floor'. However, things are changing rapidly and more and more CEOs are now taking direct interest in promoting energy efficiency and reduction of carbon footprints of their businesses. This has happened thanks to the realisation that the businesses have to become more environment friendly for long-term sustainability.

The beginning of well-structured engagement programme is made by redefining the business vision and goal. Goal has to be high enough that will energise every individual in the organisation to work for a higher-level purpose. It is during the Paris Agreement of 2015 that some of the CEOs actively participated in setting climate goals for their individual businesses. Today, that movement has snowballed to more than 630 companies setting targets consistent with the goals of the Paris Agreement through the Science Based Targets initiative.[1]

The transformation of business strategy, especially in energy-intensive manufacturing sectors, to focus on energy efficiency from an environmental perspective is indeed significant and can be approached through various means. Engaging employees at the core of this transformation is a critical component of its success.

Companies are setting clear environmental goals, often aligned with international agreements and guidelines such as the Paris Agreement or the United Nations Sustainable Development Goals. These goals typically include targets for reducing carbon emissions, water usage, waste minimisation, and improving energy efficiency. Firms are integrating sustainability and energy efficiency into their core business strategies. This involves not treating sustainability as a separate entity but as an integral part of decision-making processes, from product design to supply chain management.

Unilever, for example, developed the Sustainable Living Plan, which puts environmental progress on the household and societal levels at the centre of it's business growth strategy and charges all the company's 170,000 worldwide employees to fold sustainability into their work.[2]

Hewlett Packard Enterprise (HPE) is actively assessing the environmental impact of its operations both upstream (supply chain) and downstream (customer impact). HPE is setting ambitious energy efficiency targets for its product portfolio. By 2025, the company aims to make its products 30 times more energy-efficient compared to a 2015 baseline. Apple CEO Tim Cook has publicly committed all their products achieving 'net zero' by 2030.[3] This indicates a commitment to not only improving operational efficiency but also the environmental impact of their products.

Having an aspirational target is the beginning of the process. It must be backed up with systems and processes so that performance can be tracked on a regular basis. EnMS 50001 energy management system provides a pathway for this.

Active involvement of CEOs also ensures funding support for efficiency projects. Generally, we do not talk much about the role of funding. Track record of successful companies show that funding of efficiency programmes on a sustained basis along with aligning of the organisation with efficiency goals play the most vital role in ensuring success.[4]

DEVELOPING A SUSTAINABILITY-FOCUSED CULTURE

A sustainability-focused culture requires employees to embrace new practices and ways of thinking. Engaged employees who understand the importance of reducing emissions are more likely to champion these changes and promote a culture of sustainability. Employees are often the best source of innovative ideas for improving efficiency and reducing emissions. Engaged employees are more likely to contribute suggestions and collaborate on finding novel solutions to operational challenges. Employees who understand the rationale behind energy efficiency initiatives are more likely to support and adopt new processes. Their buy-in is essential for the successful implementation of changes in production methods and resource management. Sustainable practices require continuous improvement. Engaged employees are more committed to ongoing learning and improvement, which are vital for maintaining and optimising energy-efficient processes.

External and internal recognition and celebration of success together play a very important role in increasing the motivation and engagement level of the employee. The 'clean energy ministerial' (CEM) Energy Management Leadership Awards celebrate the leading organisations that use the global ISO 50001 energy management system standard to attain enduring improvement in energy performance. In 2022, the CEM's Energy Management Leadership Awards honoured three companies with the top Award of Excellence in Energy Management and recognising another 26 organisations with the distinguished Energy Management Insight Award.[5]

The Ministry of Power, Government of India had launched a scheme in 1991, to give national recognition through awards to industries and establishments that have taken special efforts to reduce energy consumption while maintaining their production.[6] The scheme has been designed to encourage employee participation at all levels.

Tata group an Indian MNC have many unique programmes for engaging their employees in sustainable development activities. Tata sustainability month is one such programme that provides platform to the employees to connect, adopt, and act on sustainability as individuals and collectively see the change they can bring upon themselves, their business, and the society. Organised by the Tata Sustainability Group (TSG) every June, the Tata Sustainability Month (TSM) aims to build an understanding of how the Tata group views sustainability, demonstrate how individuals and businesses can make sustainable impact and inspire their colleagues to take this understanding to the next level and make a change in their lives. Through the brand identity of Sustainable Meaningful Actions for a Responsible Tata (SMART), TSG works towards building a culture of sustainability among group companies through fresh themes and campaigns each year, developing knowledge resources and engaging activities to drive home the importance of sustainability from the individual, business, and community perspective.[7]

ENGAGEMENT PROCESS

What we understand by an engagement process? Simply put, an engagement process articulates the role of the employee in a particularly programme, her information needs and a communication strategy that will adequately meet this need. The overall process consists of few more activities as shown in Table 7.1.

TABLE 7.1
Roles and Information Need

Sl.No	Role/Responsibility	Need for Information
1	Not member of a particular project team but have roles in operation and maintenance	Company goals and strategies on energy efficiency Specifics of the programmes and projects that may impact her performance as operator
2	Member of a project team (say an external energy audit)	Customer expectation, scope of audit, audit organogram, her role in the organogram, technical capacity for performance of the role, sources of data and information, approach and methodologies for analysis and preparation of reports, schedule, budget, periodic feedback from leader on performance and customer delivery and financial performance
3	Project leader (audit, implementation)	Project goal, customer expectation, customer organisation and interface, track records on engagement, process and energy flow of the facility, technical capacities and skill competency requirement vs availability, technologies, technology vendors, financing, business performance, impact of project performance on the overall goals of her own and customer organisation

However, information and communication strategies remain the main two pillars of the engagement process. Table 7.1 shows illustrative examples of roles of employees at different levels in an energy efficiency programme and the information needs.

Assessment of the information need is followed by assessment of capacities to generate information, capacity building as per requirement and development of appropriate information and communication strategies. Having carried out the need analysis for information, we have also to assess how much to share and how frequently so that employees do not get overwhelmed with information overload. The intensity of sharing will depend upon the intensity of participation of a member in the project as illustrated by Figure 7.1.[8]

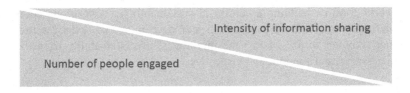

FIGURE 7.1   Employee engagement model.

At the lowest level, larger number of people are involved, say communicating to a group of employees about launch of a new programme. The communication is mostly one way focused on sharing about the need for the programme and strategy for implementation of the same. At the highest level, it is usually a multi-way process gathering and sharing of information, interacting, negotiating, taking decisions, and launching and steering the programme. Next in the process is designing a engagement protocol and launching of the same.

Digitalisation and social media have transformed the way we communicate and the impact of such communications, mostly positive but there are some negatives too. Digital tools such as email, instant messaging, and video conferencing have

FIGURE 7.2   Information and communication architecture.

made communication more accessible and efficient. This has enhanced the ability of employees to stay connected with their colleagues, supervisors, and teams, fostering a sense of belonging and collaboration. Biggest gain from digitalisation has been its contribution to skill development. Tutorials are prepared and stored in digital libraries. These can be easily accessed by the employees for undertaking training by self never worrying about the pace of learning. This empowers them to take ownership of their learning and development. This can boost engagement by providing opportunities for career advancement commensurate with developed skills.

However, digitalisation has few negative impacts too. The constant connectivity facilitated by digitalisation can lead to information overload and burnout. Employees may feel pressured to be always available, which can negatively impact their work-life balance and overall well-being.

Digitalisation brings cybersecurity risks, and employees may worry about the safety of their personal and work-related data. Concerns about data breaches can create stress and reduce trust in the workplace. Some employees may struggle to adapt to new digital tools and technologies, leading to feelings of inadequacy or frustration. This can negatively impact engagement if not addressed through training and support.

Digitalisation plan therefore, has to be carefully developed to reduce the impact of the negatives. It needs to have close co-relationship with the information and communication plan of the company. They have to move hand-in-hand for deriving maximum benefits.

CONCLUSION

In the introductory part of this chapter, the author has shared an anecdotal story on how a simple engagement strategy by a company motivated me to choose a lifetime professional career in which energy efficiency and environmental protection have always played key roles.

Employees are the most important stakeholders in energy efficiency programmes. We have highlighted how they can play both positive and negative roles depending upon the engagement strategy in any organisation. We have seen from the few case examples of CEOs, how they have taken up the leadership role for transforming their businesses driven by greater goal of sustainability. In the process, they are also developing a sustainability-focused culture across the organisations.

We have also seen that different employee engagement models are being practiced by these companies with equal amount of success in delivering the desired results. Differences are there in approach, some driven more by individual goals, some through energy management system (EnMS) and many through the reward and recognition programmes. However, the basic objective of the engagement process remains the same, defining the role of an employee in the improvement project and a communication plan that will enable her to perform the role. Thus, it can be said that there are two common pillars on which rests the engagement process, 'information' and 'communication'. The principles governing the need for information and the intensity of sharing have been explained.

The chapter has also highlighted how digitalisation is revolutionising the employee engagement process with positive impact. However, a note of caution has also been flagged about potential negative impact due to information overload and loss of privacy. The need for a structured information sharing and communication plan for energy efficiency within the overall framework of digitalisation has been accordingly underscored.

NOTES

1   *The Guardian*, Three global business leaders weigh in on how to combat the climate crisis, 2019.
2   Unilever, Planet and Society, https://www.unilever.com/planet-and-society/.
3   Apple commits to be 100 per cent carbon neutral for its supply chain and products by 2030, https://www.apple.com/in/newsroom/2020/07/.
4   From shop floor to top floor-best business practices in energy efficiency, https://www.c2es.org/wp-content/uploads/2010/03/PEW_EnergyEfficiency_FullReport.pdf.
5   ISO 50001 Energy Management System – Case Study, https://www.cleanenergyministerial.org/content/uploads/2022/09/cem-em-casestudy-3m-global.pdf.
6   https://beeindia.gov.in/en/national-energy-conservation-awards.
7   Tata sustainability month, https://www.tatasustainability.com/OurEvents/TSM.
8   IFC-<Stakeholder engagement: A good practice handbook for companies doing business in emerging markets, https://www.ifc.org/en/insights-reports/2000/publications-handbook-stakeholderengagement--wci--1319577185063.

8 Conclusion

The materials set forth in the various chapters of the book provide a sequential pathway for undertaking energy efficiency programme in industries. That does not mean that this is the only pathway for undertaking energy programmes. Large number of industrial professionals and academics continue to work extensively on the subject of energy efficiency and conservation. This implies that there is a wealth of knowledge and research available in the public domain on the subject. The contents of this book have been developed after extensive research of materials available in the public domain to ensure a differentiated value proposition for the readers of the book focusing more on 'how's' rather than 'what's' – how to make better discovery of the energy savings measures and how to implement them.

First chapter of the book has been devoted to the thermodynamics of the energy system that drives efficiency. Every part of the process, including fuel and its conversion to usable thermal, chemical, mechanical, and electrical energy has been explained in detail to underscore the sources of inefficiency and steps for reduction of the losses.

The next chapter on energy intensive industries provides brief description of processes and energy systems for these industries. The list of the industries has been chosen using few criteria such as their share of overall energy consumption in the industrial sector, potential for reduction of energy consumption and hence GHG emission, country and sector specific initiatives (global and local) on reduction of specific energy consumption and GHG emission. Sufficient information has been provided that will help readers acquire domain knowledge required for formulation of energy conservation strategies in these sectors.

The next chapter on energy audit, titled as 'Discovery of Opportunities-Energy Audit & Diagnostics' has brought out differentiated approaches that need to be considered taking into account the set objectives and target for energy savings. As against the usual practices of three types of energy audits (grade 1, grade 2, and grade 3), more different audit types have been outlined to take care of the audit goal. The concept of minimum energy design practiced in process industries have been used to develop a methodology for minimum energy audit. That has been the rationale for using a different name for energy audit 'Discovery of Opportunities-Energy Audit & Diagnostics' in this book. Diagnostics such as process synthesis, synthesis of heat exchanger networking, pinch analysis, exergy analysis etc. have been briefly discussed with a view to familiarise the readers with these high-end tools for energy system analysis. Even for conventional audit, we can use tools such as 'root cause analysis' for targeting higher levels of energy savings. A case study on diagnostics for improving efficiency of a biomass fired boiler has been included to underscore the importance of this approach. An individual organisation can exercise its own option pursuing the discovery process taking into consideration the need, required skill competency and the budget for energy audit. The book provides information

DOI: 10.1201/9781003415718-8

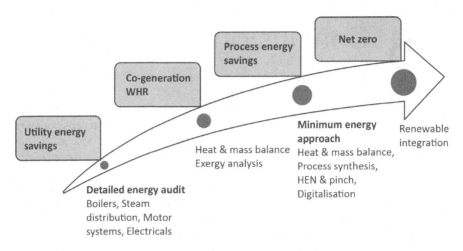

FIGURE 8.1 Net zero pathway.

and approach that can be adopted by an organisation for making its own pathway towards the 'net zero' goal as depicted in Figure 8.1. I have mentioned renewable integration in the pathway as it has a pivotal role for emission reduction, though this book is focused on energy efficiency only.

Figure 8.1 shows how an organisation can develop its own pathway beginning with the implementation of simple retrofit projects to advancing towards the 'net zero' goal.

Various modes for implementation of retrofit projects including under 'energy services performance contract (ESPC)' by energy services companies (ESCOs) have been discussed along with their merits and demerits. The details provided will help the readers in choosing the option best suited to her organisation.

Measurement and verification (M&V) of savings achieved from retrofit projects plays a crucial role in reducing risk of underperformance and improving the sustainability of savings. Approach to development of an M&V plan, various M&V options along with their applicability for different project types have and finally the structure of the M&V report post completion of the project M&V activities been elaborated.

Finally, a brief write-up on employee engagement has been included to underscore the critical role played by employees for successful implementation of energy efficiency projects and creating a conservation culture in a facility.

Index

Printed in the United States
by Baker & Taylor Publisher Services